# OUT COLD

# OUT COLD

## A CHILLING DESCENT INTO THE MACABRE, CONTROVERSIAL, LIFESAVING HISTORY OF HYPOTHERMIA

# PHIL JAEKL

**PUBLICAFFAIRS**

*New York*

PublicAffairs
Hachette Book Group
1290 Avenue of the Americas, New York, NY 10104
www.publicaffairsbooks.com
@Public_Affairs

Printed in the United States of America

First Edition: June 2021

Published by PublicAffairs, an imprint of Perseus Books, LLC, a subsidiary of Hachette Book Group, Inc. The PublicAffairs name and logo is a trademark of the Hachette Book Group.

The Hachette Speakers Bureau provides a wide range of authors for speaking events. To find out more, go to www.hachettespeakersbureau.com or call (866) 376-6591.

The publisher is not responsible for websites (or their content) that are not owned by the publisher.

Print book interior design by Jeff Williams.

Library of Congress Cataloging-in-Publication Data

Names: Jaekl, Phil, author.
Title: Out cold: a chilling descent into the macabre, controversial, lifesaving history of hypothermia / Phil Jaekl.
Description: First edition. | New York: PublicAffairs, 2021. | Includes bibliographical references and index. |
Identifiers: LCCN 2021000259 | ISBN 9781541756755 (hardcover) | ISBN 9781541756724 (ebook)
Subjects: LCSH: Hypothermia. | Cold—Therapeutic use. | Cold—Physiological effect.
Classification: LCC RC88.5.J34 2021 | DDC 615.8/329—dc23
LC record available at https://lccn.loc.gov/2021000259

ISBNs: 978-1-5417-5675-5 (hardcover), 978-1-5417-5672-4 (ebook)

LSC-C

Printing 1, 2021

Cold is immortal, unlike us. This is for all those whose lives it will extend, and to all those whose lives it has already saved. In this way cold makes us more like itself. More ceaseless. More enduring.

And for Eva-Stina, whose excitement over all the cold-related stories I've shared as I wrote this made me even more excited to cover the next topic. Now you finally do get to read the book!

# CONTENTS

# INTRODUCTION

I was out one cold January afternoon, hiking through the snow in the woods to find what I thought was an abandoned nineteenth-century farmhouse. Some months prior I had discovered it from a distance. After a couple of hours, I was tired but felt confident that it was just over the next ridge. Yet I had come to the icy waters of a creek running across my path—it appeared to be the only barrier stopping me from finally reaching my destination.

Nearby, I spotted a fallen tree that spanned the distance to the other side of the creek; heedlessly, I began crossing by balancing on its trunk. Halfway across, I lost my footing and stepped down onto the ice that partially covered the water. I broke through and sank farther as I tried to turn around to make it back to the bank. What I thought was a foot-deep waterway submerged me up to my shoulders as I struggled to get out. For the first time in my life, I felt genuinely scared. My adventure through the woods immediately turned deadly serious. I don't know how long I thrashed toward the shore. When I wasn't sinking below my nose, I tried swimming and avoiding the jagged ice around me.

When I finally emerged from the stream I knew I had to keep moving to regain warmth. I was completely soaked and dripping cold, muddy water. My hands and feet were already starting to feel numb. I followed my footprints in the snow, but by the time I reached the car I knew I was hypothermic. Luckily the drive home wasn't long. I stepped into the shower and lingered there— warmth had never felt so-o-o good. I ended up catching the worst cold I've ever experienced.

The word "hypothermia" has Greek origins meaning "under" (*hypo*) and "heat" (*therme*). Its symptoms depend on the extent of temperature decline. They initially involve shivering, movement that is poorly coordinated and laborious, and disorientation. As the condition increases, heart rate decreases significantly, and forgetfulness, confusion, and apathy set in. At extremes, victims can begin to make irrational decisions and talk incoherently. For reasons poorly understood, they've even been known to start feeling hot to the point that they take off their clothes. Before dying, they may seek confined spaces to burrow in. Many have been discovered frozen and naked.

Yet, seemingly against all logic, throughout our known history humans have also continuously sought cold for therapeutic purposes. It has been tried for conditions ranging in severity from bloody scrapes to schizophrenia and even as an indirect means for achieving immortality and time travel. Most early attempts at using cold, however, were stabs in the dark, co-opting an accessible resource that in most cases likely distracted patients from their ailments more than it actually curing them. Still, humanity has been incredibly persistent; with each new therapeutic advance, hypothermia offers a shimmering glimpse of its potential as another form of treatment.

Now, science is beginning to unlock the secrets of hypothermia in an indisputable, evidence-based manner that focuses on cold's ability to effectively slow down and suspend time before tissue damage from lack of blood flow can set in, as it often does in cases of cardiac arrest, stroke, and brain injury, for example. In this book I'll tell the story of therapeutic hypothermia—a history filled with exciting but sometimes gruesome experiments, scientists, suspended animation, head transplants, prolonged space exploration, and a host of controversial attempts at harnessing the power of cold in strange and surprising ways. This chronicle often intersects where science and fantasy meet, and where the lines between life and death are blurred. Yet based on scientific evidence accrued over millennia, we understand hypothermia better than ever before, and we have numerous new lifesaving cooling techniques at our disposal. Still, a macabre stigma hangs over the field: centuries of trying in vain to harness the power of cold have left countless dead. Current knowledge, however, is overwriting the old views. To understand how milestones in therapeutic hypothermia have been reached, the following chapters will delve into a dark history from which science is now coming out on top.

◇◇◇◇◇◇

*It already feels like an eternity since you started walking. There are no signs of human activity or even of life around here. You feel more alone than you ever have.*

*It's haunting.*

*There's no use in trying to hide your shivering if there's not even a remote chance of anyone seeing you. So you let loose, as loud as you can. "Tutttatatut chchchch ddadduududud." Indulging in audible teeth-chattering feels good, like it's going to make you warmer if you keep at it. Not only does it seem like a defense against the cold; it's actually kind of amusing.*

*You exaggerate your movements, shaking your arms around as though you were being zapped by a lightning bolt. You ponder how absurd you must look.*

*Now you actually do feel a bit warmer. That seemed to work.*

*As soon as you begin moving forward again, however, you notice just how cold your feet have become. With such little feeling in either of them, they seem more like dead weights—like they're not part of you. You're aware that a loss of feeling could mean frostbite, and you shudder to think of the consequences.*

*You try to move faster. That's always a cure for cold, you think. Got to get the blood flowing. The more you move, the better.*

*Soon, though, it seems the cold in your feet has spread to your hands. You start clenching your fists in pulses, like a heartbeat, in an attempt to make them warm. But after more pulses than you can count, your hands still don't feel warmer. You stop clenching them. The "heartbeat" stops.*

*Then, the realization of how long this journey could take hits you. It's overwhelming. You try to intentionally shiver, which seemed so effective earlier. But now it doesn't work. The shaking is not something you can actively do to keep warm; it's more like a reaction—something automatic that you can exaggerate, like a sneeze or a yawn.*

*Best to just keep moving.*

*Surely, this is going to be the journey of your life—or death.*

# 1

# MORE THAN A FEELING

## *WHAT IS COLD?*

Heat. I know it's ironic, but that's where we must begin a book about hypothermia.

With knowledge surrounding heat and temperature, we can then appreciate the strange and mysterious phenomena that happen to one's body and mind as core temperature drops. Understanding heat gives us a grip on how deadly cold can be, how paradoxical it is that cold can have lifesaving therapeutic properties, and how it has messed with the way we define life and death.

So let's begin at the beginning. Complex physics and laws of thermodynamics need not apply. In scientific terms, cold has come to be understood largely as an absence of heat rather than a property in and of itself. Heat is a form of energy that results from the motion of the constituents of matter: particles, atoms, and molecules. The more heat, or *thermal energy*, an object has, the more motion, or *kinetic energy*, is present among its atoms. When water is heated, its molecules get so riled up that an entire potful of it can move vigorously.

Temperature is basically a measurement of thermal energy; the more an object's atoms are vibrating, the higher its temperature. What's the shakiest an object can be, atomically? About 100 million million million million million degrees Celsius, according to the Standard Model of physics. The hottest temperature of anything actually recorded, however, was that of a particle collision. Although you might think it was recorded from somewhere out in space, it actually happened right here on Earth, at the Large Hadron Collider near Geneva, Switzerland. It reached a mere 5.5 trillion degrees Celsius.

And on the other end of the spectrum, what is the most static an object can be? That would be the extreme motionlessness represented by *absolute zero*, defined as a total absence of thermal energy such that an object lacks any atomic motion, save for some quantum mechanically related quiver. Absolute zero, as frigid as it gets, equates to –273.15° on the Celsius scale, which is –459.67° on the Fahrenheit scale. The lowest temperatures ever recorded have also been achieved here on Earth, in laboratories. A group of scientists cooled a sodium gas to within half a billionth of a degree of absolute zero at Massachusetts Institute of Technology (MIT) in 2003. At such low temperatures, atoms begin to exhibit spooky quantum mechanical properties. They cohere into a sort of super atom, behaving identically in terms of their movement and location.

Considering such extremes, our everyday classifications of what we think of as hot and cold seem entirely anthropocentric, arbitrary, and kinda minuscule.

Nonetheless, we, too, comprise molecules, and like all physical objects in the universe, our molecules are in motion. Thankfully not at such extreme scales. On an atomic level, when the molecules in our body begin to vibrate faster than they normally do, we feel hot, and when they vibrate more slowly than usual, we feel cold. That's as basic as it gets.

## HUMAN TEMPERATURE

As described in the introduction, the term "hypothermia" is a conjunction of the Greek roots *hypo*, meaning "lower," and *therm*, which means "heat." But it isn't as though hypothermia was always "a thing." I mean it was a thing in that people have always experienced being cold, but only recently, around 1885 as far as estimates go, has the term become commonplace.

Why did it take so long? Well, before anyone could objectively determine if a person was hypothermic, they needed some kind of reference for comparison. That is, a *normothermic* temperature, a normal body temperature, needed to be established first. Indeed, for nearly all of human existence, no one knew that cold and heat were even related to the same phenomenon: thermal energy. It was thought they were separate entities with different physical properties. Moreover, there was no way to measure heat, let alone establish a normal body temperature.

How a normothermic, stable body temperature was discovered involved some exciting experiments in the 1700s. At the time, science was largely an art of discovering what was measurable and how it could accurately be quantified. Experimentation and the creation of measuring devices were becoming all the rage, and the thermometer as a tool for measuring heat was no exception to the trend. Measurements were most relevant if they could apply directly to humans and test the limits of phenomena discovered to be quantifiable. So Charles Blagden, a British physician, and a group of his colleagues decided to perform an experiment.

On a chilly winter day, they raised the temperature in a heated room to 260°F (127°C) while taking their own temperatures at frequent intervals for as long as they could endure the heat, over multiple sessions. They published their investigations with the simple yet accurate title "Experiments and Observations in an Heated Room."

What amazed them went beyond the discovery that they could endure such heat for even an instant; no matter how many times they subjected themselves to it and how hot it got in that room, they recorded themselves as holding nearly the same internal temperature during the entire experiment, around 98°F (37°C). How was it, they wondered, that they didn't roast themselves alive? They could feel their watch chains and other metallic ornaments become unbearably hot to the touch. They observed, in astonishment, as water in a glass began to boil when they added a bit of oil to the surface to prevent it from evaporating.

Creatively and curiously, they even brought in some steak and eggs to see what would happen. As Blagden wrote in his Royal Society publication:

> We put some eggs and a beef-steak upon a tin frame, placed near the standard thermometer. . . . In about twenty minutes the eggs were taken out, roasted quite hard; and in forty-seven minutes the steak was not only dressed, but almost dry. Another beef-steak was rather overdone in thirty-three minutes. . . . The effect of the heated air was much increased by putting it in motion, we blew upon the steak with a pair of bellows, which produced a visible change on its surface, and seemed to hasten the dressing; the greatest part of it was found pretty well done in thirteen minutes.

So why didn't they themselves cook? Blagden concluded that humans have the ability to cool themselves naturally, with the aid of perspiration. And he was correct; perspiration is now well understood as a process triggered automatically by our brains, causing sweat, which removes heat upon evaporation—a quality nonexistent in inanimate tissue like steaks and eggs.

But there was something more. Blagden realized that the living, breathing human body somehow maintained its temperature in both hot *and* cold conditions. Essentially, he believed that a difference between live tissue and dead meat somehow enabled such maintenance. He established that the temperature of living things is to a remarkable extent independent of the temperature of the air around them. His writing suggests that he advocated the idea, originally credited way back to Aristotle, that such heat was a kind of innate substance, or "vital heat," possibly connected to the very soul itself.

Yet Blagden also realized that an essential property of life included the ability to heat, rather than to cool, oneself. He wrote about how animate beings maintain their vital heat in cold environments: "It seems extremely probable, that vegetables, together with the many other vital powers which they possess in common with animals, have something of this property of generating heat. I doubt, if the sudden melting of snow which falls upon grass, whilst that on the adjoining gravel walk continues for many hours unthawed, can be adequately explained on any other supposition."

These ideas were key to identifying hypothermia. Although Blagden may never have overtly proposed the concept of a normal body temperature, he laid the groundwork for finding it. Yet it wouldn't be for nearly a century that 98.6°F (37°C) was firmly established as a normothermic temperature by Carl Wunderlich, a physician in nineteenth-century Germany. At his clinic in Leipzig, he took multiple recordings of every patient's temperature over a fifteen-year span, an endeavor that generated several million data points and led him to finally arrive at this number.

But why this temperature? What is so special about 98.6°F (37°C) as a default, of all possible temperatures? If you think about it for a moment, it sounds quite high. Doesn't it? I mean, if that were the forecast for tomorrow and you had an outdoor job, you most likely wouldn't be thrilled.

Moreover, the average surface temperature of the Earth is, as of the writing of this book, 58.62°F (14.9°C). Would it not make more sense if our internal temperature were closer to that value? Instead, we require more thermal energy to achieve and maintain our temperature, and we must meet this requirement consistently *throughout our entire lifespan.* We spend enough effort and energy on all the other things we need to do to ensure our survival, like obtaining food, shelter, and mates and balancing relations with others. In this context, our normothermic temperature seems remarkably inefficient. Think about it: in tough times like droughts or ice ages, countless ancestors starved while using the energy provided by their last morsel, not necessarily for any form of purposeful action, but rather to simply maintain a seemingly inexcusably high temperature.

One reason why we have such a relatively high default temperature has to do with our bodies' enzymes: the chemical compounds that catalyze life-sustaining microscopic reactions within us. We need enzymes for thousands of processes that support digestion, respiration, muscle and nerve function, and other operations. Like anything on a molecular scale, as the temperature increases, enzymes get busier; they move around more and thereby facilitate more reactions. However, in humans they're only functional up to a maximum temperature of around 104°F (40°C). Any hotter and they will start to denature and essentially fall apart. It turns out that 98.6°F (37°C) is just right for optimizing enzymatic function, which is absolutely essential for survival.

But there's still something missing here.

Certain types of enzymes exist that can catalyze at way, way lower body temperatures than those found in mammals like us, and these enzymes still manage to enable basic life functions in the creatures that have them. We know this because countless life-forms have been discovered that thrive in outside temperatures so low they would be fatal for humans. Such organisms need their enzymes for the same life-supporting reasons we do, and unlike most mammals, they don't have fur coats to keep them warm. As an example, most of the fresh seafood in your local grocery store would never experience external heat even remotely as high as room temperature—that is, until they are being prepared for your dinner.

In fact, many Arctic species of shrimp live in water barely above freezing. Even more extreme are Antarctic ice fish that live in 28.4°F (−2°C) water. They thrive below freezing, where seawater remains in a liquid state only because of its salt content, which gives it a lower freezing point. Life-forms that can sustain and thrive in these conditions are known as *extremophiles*.

Incredibly, these extremophiles have evolved enzymes that let them live, seemingly against the laws of physics. And here we are, stuck with enzymes that require inconveniently high temperatures. Are we slaves to uniquely human enzymes? What's going on?

I know it sounds random, but according to some researchers, a lack of fungal infections, that's what. In environments and climates typically inhabited by humans, the number of fungal species that can thrive by infecting hosts drops by 6 percent for every 1.8°F (1°C) rise in temperature. Indeed, other animals with lower core temperatures are susceptible to a much wider range of infectious fungi, whereas mammals are more resistant to these forms. It seems it's no coincidence that core temperatures around 98°F (37°C) are remarkably consistent across species within the mammalian class.

As for other pathogens, like bacteria, that can and do infect higher-temperature mammals like us, heat can be used against them. A long-held theory is that a fever occurs when your body raises its temperature to combat infecting bacteria. Many forms of bacteria can't survive at fever-associated temperatures.

All of this means that we need to keep burning the food we consume as fuel in a process known as *combustion*, to keep the fire going, so to speak, in order to maintain our temperature. In fact, such an analogy, as strange as it may be, isn't too far from reality. When protein-rich, high-energy foods are actually burned, they can have an impressive output: for example, they can power large internal combustion engines. In 1938 the energy value of protein-rich dried milk was demonstrated by using it to power a locomotive for an ad campaign. Today, biofuels such as corn-based ethanol constitute the main power source for a range of vehicles. That said, even though it may be the most potent booze you've ever tried, I wouldn't consider biofuel to be a potable liquid.

The process by which enzymes help to break down food molecules and form new molecules—*metabolism*—is actually a form of combustion, the same way burning wood or gasoline is. That's because regardless of whether combustion occurs in a Formula One car or in a person, it is essentially defined, at least in part, by the release of heat. We get a good portion of our body heat from metabolism alone.

And so, the stage for this book is set. We maintain a bodily temperature of 98.6°F (37°C) because it can enable enzymes to function at peak performance while helping to keep us free from infection, all at a minimal fuel cost.

## HYPOTHERMIA AS A DEADLY CONDITION

Hypothermia, generally speaking, is of course not a good thing—a fact that should seem obvious regardless of scientific understanding. Indeed, there are clear reasons why we've evolved to find cold uncomfortable. If our bodies didn't signal that our core temperature was dropping, we could suffer deadly consequences and not realize it until too late. Severe hypothermia can cause damage to the nervous system, organ failure, and heart failure—basically, death.

Hypothermia has been known throughout human history. Before the medical term was coined and the collection of symptoms that define it were identified, it was simply known as "cold." Historically speaking, the degree (pardon the pun) to which we suffered as a result of being too cold depended largely on where we were in terms of the Earth's latitude and altitude. In northern and mountainous regions, where temperatures drop below freezing, the association between cold and discomfort and danger is inherent in cultural practices like the production of traditionally warm clothing and consumption of fat during winter.

Yet, surprisingly, hypothermia has also claimed innumerable victims in more temperate climates, even tropical ones. Victims of wet and windy conditions that make heat difficult to maintain are often caught off guard by cold, especially at night. Seafaring cultures in temperate regions have endured the loss of countless lives because of storms and accidents, both at sea and around shore. Boaters are often unprepared for the differences in temperature experienced when floating a few meters above frigid seas while far away from land. Even today, scuba divers who willfully engage with the cold water can become unsuspecting victims of hypothermia.

Since humans began keeping written accounts of events involving cold and hypothermia, we've been learning how to treat and prevent it. Much of the devastation wreaked by cold throughout the ages could have been prevented if we'd known then the science that we know now. In retrospect, many advancements seem like obvious next steps and are eye-opening in terms of how recent they are. For example, although it was commonly understood that moisture seems to have a cooling effect, this wasn't proven in regard to wet clothing until after 1950, when physiologist and mountaineer Griffith Pugh actually did the science.

Previously, under freezing conditions, clothing and sleeping bags that were saturated with moisture from sweat would often freeze. Expeditions in extreme cold could involve hours of drying time before setting out for the day. If any moisture remained, it could cause one's boots to freeze solid, increasing the risk of frostbite, which, if left untreated, could lead to gangrene and amputation. Luckily, the knowledge acquired from such events regarding cold and its effects on the human body has led to medical advancements as well as to technological advancements geared toward prevention. The invention of lighter, fast-drying materials was vital in these circumstances. Indeed, past achievements in mountaineering, such as the first Everest summit in May 1953 by Tenzing Norgay and Edmund Hillary, were largely enabled by developments that led to lighter, warmer, dryer, and wind-resistant clothing.

Yet long before all of this, for thousands of millennia, humans have thrived in harsh, freezing climates. Perhaps we've known how to deal with cold for longer than we give ourselves credit for. Is it the case that past knowledge has been lost, or usurped by commonplace but unfounded confidence in modern technology?

Fascinating evidence exists of a community that lived over ten thousand years ago on Zhokhov Island—part of Russia's northern coastal region, in the highest of the high Arctic. Signs of human inhabitants were revealed by the discovery of tools and indications of hunting activity. The archeological findings suggest that the settlement had between twenty-five and fifty permanent residents.

Further proof, much older than this, exists of human habitation in the Arctic. Scientists found evidence dating back forty-five thousand years of cut marks on a mammoth rib and a puncture-cut wound on a wolf humerus bone, both signs of human hunting and butchering. If people could survive for so long in such harsh climates, the consequences of and deaths from hypothermia throughout the world, including the modern one, seem all the more preventable and unfounded. Natural sources of clothing such as animal furs and dried grass, used for their moisture-wicking ability, can prove equally effective as some recently invented synthetic materials, if not more so.

With new technology, however, explorers increasingly regarded earlier, more primitive expertise as exactly that—primitive. It was, and still is, often thought that humans should be able to dominate or conquer nature by creating artificial advances on natural materials. Sometimes this approach works, but not always. As it applies to conquering extremely cold weather, for some explorers, this mindset failed miserably.

Perhaps such failure was never more apparent than during events surrounding polar exploration, specifically the first successful voyages to the South Pole by Norwegian Roald Amundsen in 1911, in comparison with British explorer Robert Falcon Scott's voyage there in 1912. Amundsen and his team, using knowledge gained from indigenous peoples, wore furs and traveled by skiing alongside dogs that pulled sleds carrying their gear—all traditional technologies involving traditional, natural materials, used for thousands of years. Scott and his team wore sewn clothing and attempted to use a combination of motorized sleds and ponies— all innovations in terms of polar exploration. Whereas Amundsen and his team reached the South Pole for the first time in history and traveled so successfully that they actually *gained* weight, Scott and his entire team perished. Their motorized sleds broke down, and their ponies were sacrificed because they proved unsuited for traveling long distances in snow without proper shoes. Scott and his men were left to make the voyage by pulling their gear themselves. Journalist and author Walter Sullivan, fifty years later, wrote of the events, "The Norwegians correctly estimated that dog teams could go all the way. Furthermore, they used a simple plan, based on their native skills with skis and on dog-driving methods that were tried and true."

Still, despite the right combination of traditional and innovative technology, the deadly and pervasive threat of hypothermia remains, regardless of any level of preparation.

The earliest written reference to the fatal effects of cold that I've been able to find comes from the ancient Greek historian Herodotus (484–425 BC), who wrote of the Greco-Persian wars, which started in 449 BC and ended fifty years later. He recounted the tragedy of Mardonius, a Persian general fighting the Greeks in 492 BC. According to his records, Mardonius and his fleet encountered a great storm. Three hundred ships were lost and thousands of victims drowned at sea. He wrote, "Some were . . . dashed against the rocks; and some of them did not know how to swim and perished for that cause, others again *by reason of cold.*"

After about two thousand years, soldiers were still falling victim to cold seas. Another account involving frigid waters is found in the *De nivis usu medico*, a book on the therapeutic uses of snow from 1661, written by Caspar Bartholin the Younger. Bartholin was a Danish anatomist who experienced a siege on his hometown, Copenhagen, by the Swedish army on February 11, 1659. Known as the Assault on Copenhagen, it was a victory for the Danes and led to the Treaty of Copenhagen, marking the present boundaries of Denmark, Norway, and Sweden and ending a generation of warfare.

During the attempted siege, the Danes measured the length of one of the Swedish army's moat-crossing bridges when it was left behind after a retreat. They cleverly prepared for the next attack by chipping away at the ice in the moat, which simply extended the range of the water and rendered any equal-length or shorter bridges useless. When the Swedish soldiers showed up, certainly not expecting their bridges to be too short, six hundred of them drowned in the frigid waters or died on the shores, in the snow. According to Bartholin's journals, after the attack, soldiers were found frozen, still holding active battle positions: "For some, stiffened as they were, showed the angry countenance; others the eye upraised; others the teeth exposed and threatening; some with outstretched arms menaced with the sword; others lay prostrate in various situations; and even out of the sea, when thawed in the beginning of spring, a horseman was taken entire, sitting on his horse, and holding something in his hand."

Perhaps most gripping, however, are the Napoleonic accounts of the French retreat from Moscow in 1812. Records indicate that the campaign initially consisted of over four hundred thousand French soldiers, although some estimates cite over six hundred thousand. By the time they made it to Moscow, the campaign had been reduced by about a third due to a lack of supplies and to illness. When the troops arrived, expecting an easy takeover, they were met with a largely abandoned city that had been set on fire in accordance with Russian scorched-earth tactics. After a month of consternation, Napoleon decided to retreat to France, even though winter was fast approaching and his soldiers were ill equipped to survive in cold weather. Sadly, many of them were still wearing their summer uniforms while carrying nothing to further insulate themselves. Of the retreat, Napoleon's close advisor Armand de Caulaincourt wrote a grisly journal entry:

The cold was so intense. . . . One constantly found men who, overcome by the cold, had been forced to drop out and had fallen to the ground, too weak or too numb to stand. Ought one to help them along—which practically meant carrying them? They begged one to let them alone. There were bivouacs all along the road—ought one to take them to a campfire?

Once these poor wretches fell asleep they were dead. If they resisted the craving for sleep, another passerby would help them along a little farther, thus prolonging their agony for a short while, but not saving them, for in this condition the drowsiness engendered by cold is irresistibly strong. Sleep comes inevitably, and to sleep is to die. I tried in vain to save a number of these unfortunates. The only words they uttered were to beg me, for the love of God, to go away and let them sleep. To hear them, one would have thought sleep was their salvation. Unhappily, it was a poor wretch's last wish. But at least he ceased to suffer, without pain or agony. Gratitude, and even a smile, was imprinted on his discoloured lips. What I have related about the effects of extreme cold, and of this kind of death by freezing, is based on what I saw happen to thousands of individuals. The road was covered with their corpses.

Another harrowing recollection of the deadly effect of cold on the French retreat from Moscow comes from the French physician Pierre Jean Moricheau-Beaupré as part of his book *A Treatise on the Effects and Properties of Cold*. Moricheau-Beaupré, one of Napoleon's doctors, wrote of the ravages of hypothermia on soldiers in unprecedented detail. His ability to write in so much depth, like Caulaincourt's, resulted from firsthand observations of the troops and from his own experience trying to survive the cold. His accounts of real-life events have since proven invaluable to both medical researchers and historians:

> It happened to me three or four times, to help some of those unfortunates who had just fallen and begun to dose, to rise again, and set themselves in motion, after having given them a little sweetened brandy. 'Twas in vain; they could neither advance nor support themselves, and they fell again in the same place, where they were of necessity abandoned to their unhappy lot. Their pulse was small and imperceptible; Respiration, infrequent and scarcely sensible in some, was attended in others by complaints and groans. Sometimes the eye was open, fixed, dull, wild, and the brain was seized by quiet delirium. . . . Some stammered out incoherent words; others had a reserved and convulsive laugh. In some blood flowed from nose and ears; they agitated their limbs as if groping. . . . I have observed men overpowered by cold. . . . Thus have thousands perished.

The retreat lasted months, extending into frigid temperatures and snowstorms during January and February: "Mutilations of hands and feet, loss of the nose, of an ear, weakness of sight, deafness, complete or incomplete, neuralgia, rheumatism, palsies, chronic diarrhoea, pectoral affections, recall still more strongly to those who bear such painful mementos, the horrors of this campaign."

By the end of the retreat, estimates of the remaining number of troops reach as low as ten thousand, down from several hundred thousand. The wrath of hypothermia had never taken such a toll on human life.

Even in those dark days, however, Moricheau-Beaupré was able to recognize that the human body could sustain hypothermia to a previously unknown degree. He knew of many recorded incidents of recovery from extreme hypothermic conditions:

> We read in the old "Journal de Medicine," the history of the death from cold, of a man who, in crossing the Pyrenees, was surprised by a tremendous storm, and buried beneath the snow in a state of numbness. The fifth morning he came out of his torpor; a burning thirst informed him of his existence, and made him bite the snow that surrounded him. He was quite astonished on awaking to find his tomb lighted up; he broke the layer of snow that covered his head, but his efforts to disengage himself were vain, he then implored the aid of heaven, and recalled to his soul sentiments of religion and resignation; at length persons sent in search, found him; at sight of them, the unfortunate cried out, "wine, my friends, thirst consumes me!"

Wine? Sounds like, after chilling awhile under the ice, this guy came out quite fine. After he realized that munching on snow wasn't going to cut it, his prayers were answered! One must not forget, however, that back then, wine, beer, and spirits were also sought for their warming capabilities—a belief we now know is false. Although alcohol may cause one to feel warmer, it has no actual warming properties unless served hot. In fact, alcohol increases hypothermia because it counteracts the body's natural reactions to cold. It prevents shivering, which normally causes movement and heat. It also counteracts the body's tendency to constrict the blood vessels closer to the skin so that core temperatures are maintained. Instead, alcohol dilates the vessels and keeps blood flowing close to the surface of the skin, allowing cooling to continue, if not worsen.

So Moricheau-Beaupré realized the importance of continuing to aid those with hypothermia and that attempts to revive them despite their morbid appearance and lifelessness could actually be successful. He advised that even if the patient's body feels entirely cold and appears pulseless, breathless, and unresponsive—symptoms regarded as unequivocal signs of death—they shouldn't be dismissed as dead until the "appearance of signs of general putrefication [sic]." Reflecting on such hidden vitality in hypothermia victims, he wrote, "How many resurrections might have been made on the retreat from Moscow!!!"

At that time, there was scant scientific knowledge on the physical effects of hypothermia on the body. It was simply known that victims were in need of rewarming and nourishment (more wine anyone?) and that if they had frostbite that was becoming gangrenous, the affected limb or extremity required amputation before the condition spread further. Moricheau-Beaupré's belief in ongoing effort to save seemingly lifeless victims of hypothermia set a standard of care that likely revived thousands who would have otherwise perished.

In fact, after many reports of successful revivals of hypothermic victims who upon first sight were taken for dead, the benefit of continuing treatment despite apparent death has now been confirmed by scientific evidence and solidified into the modern-day adage well-known in medical circles: "You're not dead until you are warm and dead."

## MIRACLE IN ICE

The events I'm about to describe took place on May 20, 1999, not far from my home, here in northern Norway. They concern Anna

Bågenholm, a Swedish woman who had just finished medical school at the age of twenty-nine. She took up residency in Narvik, a Norwegian town located above the Arctic Circle known for its surrounding mountain wilderness and winter sporting opportunity. Indeed, Bågenholm was motivated to apply for a residency there in part because of the seemingly endless backcountry-skiing possibilities.

Around here, May is a popular month for exactly that reason because although there is still plenty of snow, temperatures are amenable to wearing relatively light clothing, especially on days with little wind and lots of sunshine. What's more, because of the high latitude, May brings the so-called midnight sun—daylight lasts twenty-four hours, allowing for outdoor recreation around the clock.

Bågenholm, an avid skier, had set out with a couple of friends along a familiar route. After a hike up the mountain, they were ready for a thrilling descent. Soon after they started on their way down, she lost her balance on a turn and ended up going headfirst through a hole in the ice that was near a glacial mountain stream. Only her feet—with skis still attached—remained above the jagged surface. The time was 6:20 p.m.

Her friends, Marie Falkenberg and Torvind Næsheim, made a risky, frantic attempt to free her, but their efforts were in vain because of the torrent of water continuously flooding the hole and because of the thickness of the ice; they were unable to break it with their rescue shovels. They made a final, precarious attempt, in desperation, to pull her out by her skis, but it also proved useless. They decided to call for help after seven minutes of trying to free her.

The police lieutenant in Narvik, Bård Mikkalsen, received their call. In an interview with CNN after the incident, he said that he summoned a rescue helicopter from Bodø—530 kilometers south by road—but it had already left on a different assignment, to transport a sick child. He said, "You must send the helicopter to here, and you have only one minute to decide it. You have to call me back. Time is running out." The helicopter team opted to transport the child first. The rescuers then set out on their way toward Bågenholm, who had been visibly struggling under the ice the entire time, while immersed in the frigid rapids. After forty minutes, at 7:00 p.m., she stopped moving.

Ground rescue teams also sent by Mikkalsen arrived at 7:40 and were able to break the ice and free her. Leading one of the teams was Ketil Singstad, who later recalled the events, saying, "I thought we were taking a friend, dead, out of the water." Bågenholm had been under the ice by then for eighty minutes.

The helicopter arrived at 7:56 p.m. She was immediately winched into the aircraft on a stretcher and intubated so that she could be given oxygen. After the hour-long transport to the nearest hospital, in Tromsø—my hometown, 230 kilometers north by road—she was wheeled into the emergency room, where she received care from a large team of medics who were prepared and waiting anxiously for her arrival. She was extremely pale, lacked a pulse, and showed no signs of breathing. A medical report published after the ordeal evinces the severity of her condition upon arrival, stating that her pupils were widely dilated and showed no reaction to light, a subtle but sure indicator of an absence of the most basic brain activity. When connected to an electrocardiogram, she demonstrated no sign of a heartbeat. She was dead by all possible accounts.

But Mads Gilbert, the physician who headed the emergency medical team, said, "We will not declare her dead until she is warm and dead"—the now common adage.

Bågenholm's temperature at 9:52 p.m. registered at 56.6°F (13.7°C). At that point, her condition certainly fit with any notion of expiration. But slowly her blood was rewarmed, and she was given a steady supply of oxygen via a heart-lung machine.

She had now gone for at least three hours without a heartbeat of her own.

Gilbert was watching a video probe of her heart when, in disbelief and seemingly against all odds, he witnessed a contraction. Seconds later another followed, and then, incredibly, a continuous rhythm ensued. Resuscitation and rewarming efforts were immediately commenced and continued for another nine hours. After intravenous sedation was removed, Bågenholm regained consciousness and could make some movements upon instruction. It seemed miraculous that she was even alive. After twenty-eight days she was transported by air ambulance to her local hospital and then at sixty days to a rehabilitation center.

Although she suffered paralysis and required months of physiotherapy before she could regain the ability to move normally, she made a full recovery. She even resumed skiing. Today Bågenholm remembers nothing of the events in the interval between losing consciousness while trapped in the stream and waking up at the hospital. During that time, however, she set records both for surviving the longest duration without a heartbeat and for surviving the coldest body temperature ever registered.

What's more is that she came out of all of this without brain damage. Without a heartbeat to pump nutrient- and oxygen-rich blood to the brain, permanent damage can occur in as little as six minutes. Yet under hypothermic conditions, brain cells need less energy. Metabolism is slowed to a crawl, which increases the amount of time before brain damage occurs.

## HYPOTHERMIA AS A THERAPEUTIC CONDITION

Only recently have technological and medical advances enabled a detailed, accurate understanding of what happens to organs, tissues, and the human nervous system under hypothermic conditions. But cold, just as it has been known for its deadly properties, has also been known for its therapeutic power for millennia, likely longer. If you count cooling off on a hot day by taking a dip in the sea, we've been using cold to heal since well before we evolved into our present human form.

So what are the earliest written records of using cold as a therapeutic means?

Edwin Smith, an American Egyptologist, discovered them unintentionally in 1862 among the wares of a dealer in Luxor. He was intrigued by a scroll that measured about fifteen feet (about five meters) in length, on which appeared hieroglyphs that seemed to concern medical treatment. This scroll, now known as the Edwin Smith Papyrus, has been dated to around 1600 BC, but some of its text is consistent with much earlier writing, dating back as far as 3000 BC, as estimated by Egyptologist and historian James Henry Breasted.

The text, which largely concerns injury treatment, may have served as a guide for treating wounded soldiers. Breasted, who was the first to translate it into English, in 1930, conjectures that it was written by the high priest Imhotep, later exalted as the "god of medicine." Its uniqueness and value are derived from its scientific basis, especially in contrast to other papyri created around the same time, in which the origins of various medical afflictions were deemed supernatural and thought to be treatable by magical incantations and spells. Its earthly rather than supernatural approach likely exists because of the kind of injury attended to; ancient Egyptians often attributed internal injuries and diseases to supernatural causes, but not external ones. This papyrus concerns visible wounds with visible causes—surface injuries.

Historians think the papyrus was taken around 1860 by a merchant named Mustafa Agfa from a tomb in the necropolis in Thebes. Smith, fluent in Egyptian hieratic writing, immediately recognized the medical value of the text and purchased it from the dealer. The papyrus is carefully organized according to forty-eight case histories detailing the diagnosis, prognosis, and means of treatment of various injuries. It begins with those sustained to the head and then leads downward to injuries to the thorax and spine, until it ends abruptly in mid-sentence—an observation suggesting its incompleteness and the possibility that missing texts may still exist.

Case 46, the earliest known writing on therapeutic hypothermia, concerns noninfectious blistering on the chest for which a cold application is prescribed—quite specific! The application sounds equally peculiar. It consists of a preparation of "fruit, natron, and mineral, ground and bandaged on it; or calcite powder, mineral, builders mortar, and water, ground and bandaged on it."

As it contains no reference to ice or snow, the role of cold is not so obvious at first. But recall, this is Egypt, which is known for extremes in heat rather than cold. Such a mixture, prepared in this particular way, functioned essentially as a cool compress used to slow metabolism and thereby reduce tissue damage and swelling in the affected area. Its effectiveness, as far as I know, has not recently been tested, although my guess is that it had a soothing effect, similar to that of aloe vera for the treatment of sunburn. Still, the logic of the treatment surrounds the direct engagement of cold, or at least the sensation of cold.

## HIPPOCRATES FINDS THE HUMOR IN COLD

Thousands of years after the advancements made by the Egyptians, ancient Greeks began using cold as a form of treatment. It was advocated as a medical therapy by the well-known physician Hippocrates, widely claimed as the father of medicine. During his time, illness was still thought to have origins involving supernatural occurrences and temperamental gods. In many cases, people believed that disease was curable by spells and incantations. Hippocrates, however, took a more realistic, practical, and down-to-earth approach to diagnosing and treating illness, which expanded on that espoused by Imhotep and the Egyptians, to include the inner workings of the body, beyond visible, surface-level afflictions.

But that doesn't mean his theories were correct or that his practices were more effective than, say, chance. Hippocrates is regarded as an originator of humorism—the theory that healthy living equated with a balance of bodily humors. According to humorism, there were four essential fluids necessary for life: blood, yellow bile, black bile, and phlegm. Each humor was associated with an "element" (earth, fire, wind, or water) and a "quality" (wet, hot, cold, or dry). It was thought that the humors directly supported both a person's physical and spiritual essence. According to Hippocrates, "The Human body contains blood, phlegm, yellow bile and black bile. These are the things that make up its constitution and cause its pains and health. Health is primarily that state in which these constituent substances are in the correct proportion to each other, both in strength and quantity, and are well mixed. Pain occurs when one of the substances presents either a deficiency or an excess, or is separated in the body and not mixed with others."

Humorism dominated Western medical theory for over two millennia and wasn't finally written off until around the nineteenth century. Traces of it still exist in today's medical vocabulary. For example, "humoral immunity" refers to the regulation of hormones and antibodies throughout the body. The theory was, however, conceived before any notion of heat or cold was based in physics and confirmed by scientific evidence, let alone understood in modern medical terms. To provide some context: at that time, in ancient Greece, people didn't understand that hot and cold were actually of the same nature. Rather, it was thought they were independent entities, each with unique properties.

So, it was all about balance. An individual enjoyed good health when humors and qualities were each present in the right amount. Heat was considered a quality of yellow bile, and cold was associated with phlegm. Even though these qualities maintained an equilibrium during a healthy state, hot and cold were in opposition, working against each other. Some ancient Greek texts use political and military vocabulary to describe their combative relationship. But although the "battle" between hot and cold raged within one's corporeal self, their mixture was not associated with discomfort or pain. Rather, it was when the two states separated that one felt ill. Furthermore, whichever quality had left the body would often naturally return to restore the balance. In most cases, there was no need for medical attention. For example, if a person became chilled from, say, falling in cold water, heat would naturally be restored within their body upon their entering a heated room.

Charles Blagden would have rolled his eyes.

Heat was often identified with swelling and as a means of treatment, and Hippocrates suggested using various cold applications. He claimed that cold was useful in treating pain: "Swelling and pain in the joints unassociated with ulceration, gout and spasms, are mostly relieved and reduced by cold douches and the pain thus dispelled. A moderate numbness relieves pain."

Cold, he argued, could also be used to treat tetanus. He wrote, "Cold affusion serves to recall the absent heat, and thereby, terminate the disease." Again, quite specific and perhaps, at first glance, rather obscure! Nonetheless, according to a 2017 study in *The Lancet*, a prestigious medical journal, one million cases of tetanus are estimated to occur worldwide each year, with more than two hundred thousand deaths. Although cold is not currently regarded as an effective treatment for tetanus, as a bacterial infection it can be made worse by excessive heat. Hippocrates was at least on to *something*.

Finally, in more practical terms, he thought that pain from tumors and in the joints could be relieved by the "sedative" properties of cold. Indeed, cold was one of the earliest forms of anesthetic, easily obtainable in any place with snow, ice, or even a cold body of water nearby.

## GALEN'S COOL

Galen of Pergamon (AD 129–210) advanced the practice of cold, though still in the context of humorism, before the age of thermometers. His influence in the story of cold is even felt today. Although he was Greek, he served for a period as the Roman Empire's primary physician. His medical research has been so influential that his theories were a major influence on European medicine for the next fifteen hundred years. He was such a prolific writer that his texts comprise the largest body of work of any ancient Greek scholar.

Humorism had a significant influence on Galen's theories. At that time, fever was thought to be its own illness rather than a symptom of others, and it was believed to be caused by an excess of heat. Thus, he argued that the best treatment for strong fevers was cold affusion, or the application of cold to the body.

Crucial aspects of his specific regime live on today. You're likely to be familiar with the phrase "cool as a cucumber." Galen advocated eating fruits such as cucumbers or watermelons for their cooling quality. The association holds today for these popular summertime cocktail and salad ingredients. Their cooling effects originate from their relatively high water content; water, being denser than air, will remain cool as the day heats up.

Galen is also the inventor of "cold cream"—yes, the kind you can buy at your local pharmacy. During his time, *ceratum refrigerans* was a simple mixture, likely in the form of beeswax and water, to which aromatic softening elements like rose petals could be added. Also intended for treatment of fever, the emulsion was sought for the cooling sensation it produced on the skin when it dried. Currently, cold cream, or *cérat de Galien* (Galen's wax), as it is known in France, has a different composition and is valued for its antiaging, moisturizing properties and its effectiveness as a makeup remover and skin cleanser.

## AVICENNA AND THE CHILLY "STUPEFACIENT"

Around the turn of the first millennium, Avicenna—also known as Ibn Sina, Abu Ali Sina, and Pur Sina—advanced the practice of using cold by advocating it as a means of treatment for pain and for, ironically, injury from cold. An astute polymath in the Golden Age of Islam, Avicenna wrote mainly about philosophy and medicine. His most comprehensive work on the latter was published as an encyclopedia, appropriately called the *Canon of Medicine*. It contains references to using cold in the form of snow or ice water as an anesthetic. Specifically, he refers to it as a "stupefacient": "The most powerful of the stupefacients is opium. Less powerful are: seeds and root-bark of mandrake; poppy; hemlock; white and black hyoscyamus; deadly nightshade; lettuce-seed; snow and ice-cold water."

I'm inclined to think that the placement of "snow and ice-cold water" at the end of the list implies that they should be used as last resorts relative to something as potentially powerful as opium.

Avicenna also provides a diagnosis, prognosis, and means of treatment for serious frostbite, surprisingly involving snowy water among other things:

> For you know that once the (freezing) cold penetrates into a member, not only is the innate heat extinguished, but the very substance on which that heat depends is destroyed. The tissues are then at the mercy of putrefaction. So there is an urgent need. . . . The best thing is to place the limbs in snow water, or into water in which figs have been boiled, or cabbage,

or myrtle (i.e., odoriferous things), or into dill water, or chamomile water. All these are beneficial. A good local application is made with pennyroyal. Wormwood of Pontus, and betony, and turnip are also good medicaments for the purpose.

Although such means of treatment for frostbite may now appear comical, Avicenna does not suggest what might seem an obvious emergency measure: placement next to a fire or immersion in hot water to regain heat as soon as possible. This is probably not coincidental, as rapidly heating a frostbitten extremity can cause further damage; cells close to the surface are thawed before blood vessels can reopen and carry oxygen to them, causing them to die. Avicenna was unaware of this phenomenon, but during his era people had observed that quickly thawing frozen food—fruit, for example—would cause it to lose consistency, become flaccid, and "putrefy" faster than if thawed in cold water. Accordingly, his writings incorporate this conceptualization into his suggestion for frostbite treatment: "Some treat frostbitten parts with great advantage by plunging them into very cold water in the same manner as is usually done with fruit which has been frozen." Perhaps he was also aware that quickly reheating frostbite causes considerably more pain than a slow rewarming. Current practice suggests using tepid or warm water, no more than 104°F (40°C), to thaw frostbitten body parts and allow blood to recirculate.

That said, although it was known then (as it is now) that applying too much heat to a frostbitten body part is harmful, back then—and this can't be overstated—there was *no conceptual understanding of heat as thermal energy*. The idea of understanding or measuring heat objectively, as temperature, simply did not apply to medical practice. Although Avicenna's beliefs were on the right track, he was essentially blind in terms of what was actually happening to his patients on any fundamental level involving microscopic effects of cold and heat on human tissue.

And so, during Avicenna's time and for over a century after his death, although humorism was dogma, little effort had been made to actually quantify humoral balance or its associated properties of heat, cold, moisture, or dryness. Entirely subjective methods of assessment were the norm. When it came to measuring a person's "heat"—for example, that of someone who appeared to have a fever—it was time to break out a tried and true instrument: the hand, the go-to "thermometer" of the day. In fact, "hand thermometry" was an art that took practice and is likely still achieved by many modern-day parents. As late as 1871, a medical textbook written by E. Segui describes it: "The hand was not simply placed on the skin surface but remain[ed] hovering a while above or near, to perceive the heat exhaled therefrom; then enter[ed] upon a slight contact with this surface to receive the impression of its most superficial temperature; then by a firmer pressure receive[d] the full impression of the skin's temperature; then by gradually deeper pressures acquire[d] the impressions of the deeper seated combustions."

To conclude, the subjective, qualitative views of heat and of cold that I've summarized in this chapter dominated Western medicinal theory for millennia, without any major departure from the occult beliefs and practices surrounding humorism. Thus, other than using cold to restore balance to the humors and as an anesthetic, it didn't merit much regard. Still, the fact that it was considered at all by such influential thinkers and that it found its way into their texts is nothing to sneeze at.

Avicenna, Galen, Hippocrates, and the Egyptians, without any understanding of heat and cold as physical, thermal properties, clearly had some idea that cold, despite its usual adversity, appeared helpful for treating certain medical conditions.

◇◇◇◇◇◇

*Suddenly it hits you: you've completely lost track of time. Just how long has it been? You realize you'd rather not know. You're starting to feel colder than any frigidness you have ever experienced.*

*As you scan your surroundings, you find it increasingly difficult to focus on anything.*

*You keep glimpsing movement in your periphery, but when you look to discover what's there, you see nothing. Out here, in such unfamiliar territory, the experience is spooky—like you're being teased by a ghost that keeps disappearing and reappearing. Perhaps something is really there, but you just can't seem to focus on it.*

*It's time to stop again and rest. Resting feels glorious, a moment of peace and comfort that you take some time to revel in. You never want this feeling to end.*

*In fact, despite your dire circumstances, a distinct numbness clouds your thoughts and emotions. Although there are moments when you feel the combined weight of all the relevant torments, you become more and more detached as time ticks on.*

*With considerable effort, you get up and resume pressing ahead. Your motivation, however, is becoming as weak as you are. You're confident that you're simply heading, in a robotic fashion, into oblivion with each aching step.*

# 2

# THE INVENTION
# OF THE THERMOMETER

## COLD BECOMES A SCIENCE

Our understanding of what heat is and that it can be measured, like many things scientific, evolved in slow incremental steps. These advancements yielded key innovations, such as the notion supported by the theory of bodily humors that health problems originate physically and not supernaturally, and that they can be prevented or treated, also by physical means. Still, there remained no fundamental understanding of cold or any concept of hypothermia, and so experiments involving cold for its potential to heal comprised daring schemes that garnered just as much praise from their advocates as criticism from detractors. Although many of these exploits ultimately helped to ground cold therapy in science, others propelled cold therapy into dangerous and deadly territory.

During Avicenna's time, and for centuries after, the spread of information was still relatively slow. People had limited access to published material, and world literacy was still far below

50 percent. Even if literate persons gained access to medical texts, chances are they were written in Latin or in another foreign language, either of which most people were unfamiliar with. Word of mouth was still the primary highway on which information traveled, and therefore any scientifically oriented communication about cold likely existed only as a conversational topic, open to mutation and distortion, just as in a game of telephone.

In the ages before science, many healing practices continued to be dominated by absurd lines of reasoning: mercury for the treatment of influenza, arsenic for syphilis, and even the external application of human excrement for the common cold. Such traditions set the stage for some truly misguided and wacky approaches to harnessing the potential seen in cold, which were centered mainly around cold water as a cure for everything from measles to smallpox to insanity.

## HEAT EXCHANGE:
## FROM MAGIC TO MEASURABLE PHENOMENON

By the early seventeenth century, the polymath and popular Tuscan genius Galileo Galilei was forging new, innovative, and empirically based methods in astronomy, physics, and engineering. Through experimentation, he was able to discover physical laws that remain relevant today, like the law of falling bodies: all objects fall at the same rate, regardless of their weight. His approach, which involved using evidence, mathematical calculations, and quantifications, demonstrated an understanding of the natural world that was as advanced as it was accurate.

Aside from his better-known astronomical discoveries, like the rings of Saturn and the existence of sunspots, Galileo is credited with the invention of the thermoscope, a device for gauging heat. It's not the same as a thermometer, though. It couldn't measure—meter—temperature because it had no scale. It was also quite inaccurate because it wasn't sealed and was therefore biased by changes in barometric pressure in the atmosphere. Ultimately, the thermoscope didn't offer a major innovation over any skilled person just using their hand.

Nonetheless, this was the dawn of the age of the Scientific Revolution, which began around the mid-1500s with Copernicus's famous treatise, *On the Revolutions of the Heavenly Spheres*. A growing support for empiricism meant that the process of measurement was becoming increasingly central to gaining objective insight into the natural world. What was found to be measurable and quantifiable marked the frontiers of discovery. During this era, a flurry of measuring devices and units of measurement were invented, eventually forging the standard units we have in place today. Indeed, Galileo's scientific innovation was an essential context for the current understanding of temperature.

Then, around 1612, with a name so nice he used it twice, Santorio Santorio, another Venetian scholar and associate of Galileo, made crucial conceptual advances to the thermoscope. He's been credited with adding a scale—an advancement about as fundamental as the invention of the device itself. The early thermoscopes basically consisted of a vertically oriented glass tube with a bulb at the top and a base suspended in a pool of liquid such as water, which ran up a length of the column. As the temperature of the air in the bulb increased, its expansion changed the height of the liquid in the column. Santorio apparently placed a calibrated scale on the glass. His writings indicate that he set the maximum by heating the thermoscope's bulb with a candle flame, and he set the minimum by contacting it with melting snow.

Furthermore, Santorio may have been the first to apply the thermometer to the field of medicine, as a device for objectively comparing body temperatures. To take a measurement, the patient would either hold the bulb with their hand or breathe on it. Santorio's advocacy of the thermometer as a medical instrument that could allow for absolute, rather than relative, temperature measurement is found in a passage of his 1612 commentary, *Commentaria in Artem Medicinalem*:

> Furthermore, both Avicenna and Galen claim . . . that our sense of touch is the judge of all [types] of heat: if the species of heat were different, the touch would not be the right judge of them. Indeed . . . he [Galen] assigns to the touch the judgment about the equality of heat in children and young men[.] Galen urges us to touch many and different objects, that is to say water at first not too hot and temperate, then the very limbs [of the body] yet according to this rule, which consists in comparing the weak to the weak, the stocky to the stocky, the fat to the fat and not the unexercised people to those at rest or those fasting to those who are full. This way of measuring the degree of heat is certainly misleading. As for our part, we resort to the glass instruments . . . which surely cannot mislead us. By means of these instruments we have tested whether heat is the same in children and young men. The experiment consists in placing the hand of a child and then of a young man on the glass bulb of the instrument for an equal interval of time; from this we understood that the water descent was the same in both ages which means an equality of heat.

Here, another crucial observation regarding the advancement of science, and therein medicine, can be inferred: Santorio implied that it was necessary to compare temperature measurements over a wide sample of individuals. In doing so, he seems to have discovered that heat does not come in different forms or "species," as was commonly thought, but rather that it comes in different amounts or "degrees."

In the 1650s, another breakthrough occurred when Ferdinando II de' Medici, Grand Duke of Tuscany, made key design changes to the old thermoscope. De' Medici is cited as the first to create a

sealed design, unaffected by air pressure. His thermoscope consisted of a vertical glass tube filled with "spirit of wine"—distilled
wine—in which glass bubbles of varying levels of air pressure rose
and fell with changes in temperature. He was so into measuring
heat that in 1657 he started a private academy, the Accademia del
Cimento, where investigators explored various forms and shapes
for their thermoscopes, including ornate-looking designs with
spiraling cylindrical columns. The devices made for some impressive feats of glassblowing and were noted for their craftsmanship
and their functionality in greenhouses, which were becoming an
aristocratic trend. Owing to the improvements in both the form
and function of the instruments, their demand grew steadily over
the final fifty years of the seventeenth century, when they became
known as "Florentine thermoscopes."

Yet even with this improved functionality, accurate temperature measurement had quite a ways to go. There was still no accepted standard for calibration. The ways in which people tried
to find a reference point were ridiculously arbitrary; they used
standards as wide-ranging as the melting point of butter, the internal temperature of animals, the cellar temperature of the Paris
observatory, the warmest or coldest day of the year in various
cities, and "glowing coals in the kitchen fire." And although the
instruments technically had measurement scales, the intervals of
those scales were also completely arbitrary. They were often sim-
plistically marked with equally spaced dots, with perhaps every
fifth or tenth dot being a different color, to imply some form of
unit, although no standard actually existed. No two thermometers registered the same temperature. It was a mess.

Enter Danish astronomer Olaus Rømer (also known as Ole
Rømer), who heralded an innovation that would change thermometry forever. In 1701 he had the idea to calibrate a scale relative to something much more accessible: the freezing and boiling
points of water. Similar to the way we measure minutes within
an hour, the range could be divided between these points into 60

degrees. Although this is what he could have done, and it would have been great, he didn't quite get there. Awkwardly, since he had originally used frozen brine as the lower-end calibration point, his measurement of the freezing point of water occurred at 7.5 degrees. Rømer's final scale, between water's freezing and boiling points, began at 7.5, running up 52.5 degrees to 60. Perhaps unsurprisingly, his scale is known today as the Rømer scale, and although it bears historical importance, it is not formally in use anywhere.

As interest for thermoscopes continued to grow throughout Europe, a young merchant who would be key in their development discovered that the implements were becoming an increasingly popular trading commodity. He also found them to be utterly fascinating. His name was Daniel Gabriel Fahrenheit. Although you're probably not surprised to hear his name come up here, his story is a rather remarkable one.

Fahrenheit was born in Danzig, Poland (now Gdańsk). He came from a successful mercantile family that operated a trading branch in Amsterdam, where the building of greenhouses was on the rise, increasing the demand for thermoscopes. In 1701, when Fahrenheit was only twelve, both his father, who was a merchant, and his mother met an obscure fate: they died from eating poisonous mushrooms. Along with his siblings, he was taken in by new guardians and appointed to a merchant's apprenticeship. Young Daniel, however, didn't care much for the profession. He was more interested in science and glassblowing (you can see where this is going). Studying, creating, and designing thermometers and barometers became his calling. But in his relentless pursuit of these activities, he accrued debt that he was unable to cover.

Although Fahrenheit was entitled to an inheritance from his parents, he could not yet use it to pay the debt. Instead, his new guardians were held responsible for it. Their solution: appoint him as a seafaring laborer for the Dutch East Indies Company so that he could earn the money to repay them. Fahrenheit escaped his fate by fleeing the country. He needed to wait out the years until he was twenty-four, when he would be entitled to his inheritance and able to make good on his financial obligations. So he wandered through Germany, Denmark, and Sweden for twelve years while continuing to pursue his love of science.

Eventually, his path crossed with Rømer's, and he settled in Amsterdam, where Rømer lived. Their collaboration spawned the first quicksilver (mercury) thermometer, which afforded greater accuracy and precision than its predecessors. It was around this time that Fahrenheit became known for the design and construction of thermometers that reflected his glassblowing skills. And finally, he was able to make multiple thermometers that gave consistent readings because of the improved design involving mercury.

With the rising demand for thermometers, Fahrenheit was in the perfect position to develop his eponymous scale. He based it on Rømer's but calibrated the zero point to the freezing temperature of a brine solution made of an equal mixture of water, salt, and ice—substances accessible to all. He found that the surface of a solution of equal parts water and ice froze at 32 degrees, which is now the commonly known "freezing point" in the Fahrenheit scale. With two more increments of 32—that is, 96 degrees—the scale matched what Fahrenheit measured as the temperature of the human body, as calibrated by placement of the thermometer under his armpit. It all fit nicely together, and so the gauge caught on, eventually becoming temperature's first standard scale.

With 32 a seemingly arbitrary number for the base of the thermometer, it might not be entirely surprising that his choice has served as fodder for conspiracy theorists. There were rumors that Fahrenheit was an active Freemason and based his scale's starting point on the "32 degrees of enlightenment," which accord to some rites of Freemasonry. However, no official records of his membership in the Freemasons exist. Meanwhile, arbitrary but significant numbers were low-hanging fruit for absurd hypotheses. For example, thirty-two is also the number of

* piano sonatas by Ludwig van Beethoven;
* Kabbalistic Paths of Wisdom;
* black or white squares on a chessboard, and the total number of pieces (black and white) at the beginning of the game;
* and teeth in an adult human.

When one considers the sensibility actually put forth by Fahrenheit to mark the scale in 32-degree increments between the freezing of the brine mixture, of the water and ice mixture, and his (almost accurate) measurement of human body temperature, the gauge doesn't seem so arbitrary. Its oddness was highlighted only after the world began to adopt the metric system, which sets the temperature range between freezing and boiling at 0 to 100. The Fahrenheit scale still exists today in a limited part of the world that includes the United States, Liberia, Saint Kitts and Nevis, and Palau.

As Britain established colonies over an expanding region around the globe, the convenience of the metric scale and its application to a variety of measurements—currently: distance, volume, mass, electricity—meant that more and more colonies adopted it as a standard for trade. Because it integrated with the universal numeric system in units of ten, commonly known as "base ten," it was intuitive and made calculations easier.

By the mid-twentieth century, the metric system dominated the globe. Yet, strangely, the anglophone world still hung on to the Fahrenheit scale. The UK finally went metric in 1965 along with other commonwealth countries like Canada, Australia, and South Africa. In the United States, Congress passed the Metric Conversion Act in 1975, which made metrication voluntary—

a move that turned into an Achilles' heel. Given the choice to switch to a new system or stick with what had worked for them all their lives, nobody wanted to change. The idea of converting to the metric system was met with staunch opposition. The act fell flat. Today, the case for metric conversion is motivated by better international scientific cooperation, an easier, less error-prone system of measurement, and the economic benefit of not having to make products with separate scales—one for international shipment and one for domestic trade.

Upon the worldwide adoption of the metric scale, the Fahrenheit system was succeeded by the scale invented by Anders Celsius in 1742. Celsius was a Swedish astronomer, born in Uppsala in 1701.

What he did was also essential to thermometry. He made the calibration process more accurate by simply using the freezing and boiling points of water at sea level—no more salt mixture requiring its own measurements, à la Fahrenheit. On his original scale, however, 100 degrees was the *freezing* point. Rather than Celsius, it was Jean-Pierre Christin, a French physicist, mathematician, astronomer, and musician, who, around the same time as Celsius's innovation, conceived of a similar arrangement but with 100 as the boiling point—the current Celsius scale.

So there we have it: temperature measurement as it is today. With functional thermometers that were finally able to accurately measure body temperature, you'd think that before long the medical community would establish a normothermic measurement.

Right?

Nope. Despite the invention of the thermometer, the concept of heat was not *yet* fully understood. I know—science can be slow and slogging like that.

Although major advancements and conceptual changes in physics were about to take hold, especially those sprung from the genius of Isaac Newton, incredibly, hot and cold were still widely thought of as separate entities—like apples and oranges. And to give you some sense of the progress in the world's understanding of these properties, they were still largely shaped by the theory of humorism.

Also, thermometers remained relatively rare and expensive. Medical thermometry mostly advanced conceptions of the relation between heat and fever. Dutch physician Gerard L. B. Van Swieten (1700–1772), along with many of his contemporaries, observed that illness, especially fever, correlated with changes in temperature, as measured with thermometers. Van Swieten is quoted as saying, "When such a thermometer, first used on a healthy man, and marked accordingly on the scale, is either held in the hand of a fever patient, or the bulb placed in his mouth, or laid on his bare chest, or in his axillae [armpits] . . . the ascent of the mercury to different elevations will show how far the fever heat exceeds the natural and healthy."

These advancements were essential for developing therapeutic notions of hypothermia, and they set the stage for a flurry of discovery. But, again, not anytime soon. It would be at least a hundred years before the invention of the conveniently sized clinical thermometer that is in use today.

## HYDRO *THERAPY?*

Until this point in the history of using cold for therapeutic purposes, the practice was nearly entirely constrained to local applications of cold mediums—to reduce swelling, for example. Yet around the turn of the eighteenth century, insight into natural phenomena was increasingly achieved through experimentation and invention. And so a pattern began to emerge that got people talking: cold-water bathing seemed to lead to health benefits, and it could also feel great—at least afterward.

One of the first to realize this connection and write about it was John Floyer (1649–1734), an eccentric English physician whose primary claim to fame involved the practice of measuring

42

pulse rate. In fact, he designed a special kind of watch just for that purpose. Floyer had a fervent aptitude for experimentation and an insatiable curiosity. He was known for experimenting on himself and for using the medications that he prescribed to his patients, calling out other physicians "who caused their patients to swallow what they dared not taste themselves." In fact, apparently Floyer advocated prescribing medicine based on diagnoses obtained by tasting and smelling his patients' bodily humors: "The agreements and contrarities betwixt the tastes of the humors of the body and the taste of the medicine, it was easier for the physician to infer that by a medicine of the same taste, the humors of the body might be preserved."

Floyer traveled throughout his native England, visiting swimming holes where locals cooled off regularly. After speaking with them, he accumulated records of bathers' behaviors, routines, and comments that suggested their habits extended beyond simple cooling purposes. Using these records, he wrote a comprehensive manuscript titled *An Enquiry into the Right Use and Abuses of the Hot, Cold and Temperate Baths in England*. It was published in 1697.

But what exactly were his insights after such a *tour de bath* (which, by the way, included the actual town of Bath and its famous Roman-period pools)? Bathing wasn't a frequent or common activity at the time; water was seen as somewhat of a threat because most people didn't know how to swim, and drowning was relatively common among seafarers and dockside workers. Not to mention that the idea of a bathroom shower hadn't even been conceived of and would likely have been unappealing, especially during cooler months. Why would anyone *want* to get rained on?

What Floyer discovered was that cold bathing was beneficial for "personal hygiene" and general well-being. He wrote, "The benefit of bathing in rivers is very great. . . . The general effects which experience assures us that it produces are to cool in the Dog-days, to cleanse and moisten the skin; it cures thirst, causes sleep; produces much urine, prevents fevers and feeds thin bodies and creates in them an appetite, and helps their digestion."

By the arrival of the mid-eighteenth century, cold-water bathing was recommended by another notable physician, William Cullen, of Scotland. Cullen was an agriculturalist and a leading professor at the new Edinburgh Medical School. Indeed, defined by his prolific and insightful scientific and intellectual output, he was a central figure in the Scottish Enlightenment. Details of Cullen's medical knowledge and advice are contained in hundreds of letters to his patients and their caretakers, many of which still exist today. Advancing Floyer's broad conclusions about bathing simply being good for you, Cullen actually prescribed it as a treatment.

Take, for example, the regimen he advocated for Prince Rezzonico of Rome, a senator and a nephew of Pope Clement XIII. The prince was suffering from gout. In a letter written in 1786 to the physicians caring for Rezzonico, Cullen recommends a cold-water routine: "When there is the least appearance in gouty dispositions of a determination to the extremities I abstain from cold bathing, and often from the beginning of Spring to the beginning of Summer when such determinations most commonly take place I also abstain from cold bathing, but in other circumstances and in other Seasons I have found it of great service to gouty persons, and especially in Flaccid and inert constitutions, such as I think our patient is to be."

He goes on to suggest using what he refers to as a "shower bath"—a tub suspended by ropes and pulleys from which water was poured onto the patient:

> You must practice only with the Shower bath for the constitution of which you must be directed by some persons who have been acquainted with it in Scotland, or if you cannot find such persons to direct the making of it at London let me know and I shall send you very quickly the chief part of the apparatus ready made from hence. . . .

For the first day to the four gallons of cold water you are to add one gallon of boiling water, and stirring there a little together, they are to be immediately let fall upon the Patients head and Shoulders, and when it has now quite down he is to be immediately dried all over with coarse towels, and have his ordinary Cloaths put on. This Shower of water, the person will hardly feel cold, or very moderately so, and cannot call it a painful practice. But every day after wards half a pint of cold water is to be added and half a pint of boiling water is to be kept out and thus in the course of fourteen days a person is brought gradually to bear the water quite cold, and when more cold water may by degrees be added to the Cylinder till it is quite filled.

No physician around that time, however, advocated cold therapy more vociferously than James Currie. Currie, a student of Cullen, is well-known for his experimentation with cold water and for his advocacy of using it to treat fever. He was motivated by his own experiences with a recurring condition and the positive outcome of his self-experimentation with cold water.

As a youth, Currie held a keen interest for travel and discovery, and in 1771 he sailed to Virginia on a merchant ship. It didn't take long for him to decide, however, that life on the new frontier wasn't all it had been made out to be. He soon took more of an interest in literature than in merchant employment. He also began to suffer from fevers that compromised him physically. In 1776 he undertook the study of medicine, attending the now esteemed medical school in Edinburgh where Cullen was a professor.

There, one of his early research interests concerned cold. Currie was inspired by a traumatic event that happened upon his return to Scotland: after an intense storm, he witnessed some of the victims of the wreck of an American ship perish in the Irish Channel. He watched sailors struggle for their lives in the frigid waters surrounding the ship; others had climbed onto the capsized vessel to await rescue. To his bewilderment, the survivors were the ones who had remained immersed in the cold water, whereas those who perished were the ones who had awaited rescue while atop the overturned ship. The experience would make a lasting impression and would become a pillar on which his future as a physician and researcher would stand.

After personally interviewing the survivors, asking them to detail their accounts of the tragedy, Currie conducted controversial experiments in an attempt to understand what had happened. As best he could, he simulated the conditions of the disaster by immersing subjects in similarly cold water and weather. He then took careful measurements of their temperature. Although he participated in the studies himself, it is unlikely he subjected himself to a drop in bodily temperature to 88°F (31°C)—the lowest he recorded in his subjects. How he recruited these brave—or should I say unfortunate—subjects remains somewhat of a mystery.

Through Currie's diligent application of thermometry, he discovered *afterdrop*—the phenomenon wherein a hypothermic body's core temperature cools down further upon rewarming. Afterdrop results from the reopening of blood vessels that permeate chilled tissue and the subsequent cooling effect of blood circulated through these cold vessels.

His experiments also yielded some of the first scientific research into the effects of whole-body cooling and introduced the now-common method of placing a thermometer under the tongue for an accurate estimate of a patient's core body temperature. He found this approach better than previous methods of taking one's temperature in the hand or under the arm.

Currie applied the knowledge he gained from these inquiries toward an investigation of body heat and sweating. He came to a conclusion similar to that discerned by Charles Blagden, who you may recall had conducted hot-room experiments and discovered that the body cools itself by sweating. Currie concluded that the purpose of sweating was to aid in cooling, rather than to eliminate toxins or excess bodily humors, which was the commonly held belief at the time. He connected this insight with the well-known observation—and his own experience with illness—that patients suffering from a high fever often perspire: he surmised that cooling could revolutionize the treatment of fever!

He was deeply inspired by the experience of William Wright, a fellow Scottish physician. Wright had, in 1777, contracted a fever from a sailor (likely typhoid) while near Jamaica and treated himself by having bucketfuls of cold water flung upon him at regular intervals. Of his experience he wrote:

> September 9th, having given the necessary directions, about three o'clock in the afternoon I stripped off all my cloaths, and threw a sea cloak loosely about me till I got upon deck, when the cloak also was laid aside: three buckets full of cold salt water were then thrown at once on me; the shock was great, but I felt immediate relief. The head-ache and other pains instantly abated, and a fine glow and diaphoresis succeeded. Towards evening, however, the febrile symptoms threatened return, and I had recourse again to the same method, as before, with the same good effect. I now took food with an appetite, and, for the first time, had a sound night's rest.

Wright continued his cold treatment for two more days and experienced no relapse. And although he advocated cold-water treatment, he had not taken a single temperature measurement to confirm any relation between his experience and his actual body temperature. Currie was now perfectly poised to validate the effect quantitatively by measuring the temperatures of feverish patients while treating them with cold.

That said, Currie did not simply and crudely want to hurl frigid water on the first fever sufferer he met. He started by taking measurements of a patient's temperature at regular intervals, beginning as close to the onset of fever as possible, to determine its course. He also made a general assessment of the patient's condition to make sure they wouldn't be overwhelmed by the cold-water treatment.

At that time, fever was attributed to a "spasm" of activity in the blood vessels; at its onset the spasm caused the patient to become pale and cold to the touch and then to become hot afterward as a result of dissipating heat, which was released by sweating. Specifically, Currie writes of the effect of cold water: "The sudden, general, and powerful stimulus given to the system [by the cold water], dissolves the spasm on the extreme vessels of the surface, and of the various cavities of the body: the sudden and general evaporation carries off a large portion of the morbid heat accumulated under the skin; and the healthy action of the capillaries and exhalents being restored, the remaining superfluous heat passes off by sensible and insensible perspiration."

Currie set out to carefully initiate the treatment at the point when a patient's fever was high but dropping, as part of its natural course. Thus, he targeted the cold treatment to be administered just after the patient's temperature had peaked—when they had stopped shivering and no longer felt cold, but had not yet begun to feel hot and to start sweating. Through his experiences with patients he designed careful cooling regimens dictated by diligent temperature measurements to pinpoint these moments. His regimens often required patients to have cold water poured over them, repeatedly, after a set number of hours. Yet despite this scientific approach, Currie was unable to determine a precise methodology and rather owed much to intuition.

His essays, composed around the turn of the nineteenth century, were well circulated and cultivated an initial enthusiasm among doctors and caregivers who heralded his innovativeness. They were relatively accessible, having been written in English rather than Latin, the academic standard of the era, and were largely free of jargon and technical terminology. After four editions of writings on cold therapy, which were initially published in 1797 as *Medical Reports on the Effects of Water, Cold and Warm, as a Remedy in Fevers and Other Diseases*, his method was extended to successful treatment of scarlatina (scarlet fever), smallpox, measles, influenza, shipboard fevers, and tropical fevers. Currie himself practiced what he preached, as documented by his writings concerning the treatment of his own sons for scarlatina: "I shut myself up with these boys; and with plenty of pump water and a pocket thermometer, prepared, not without anxiety, to combat this formidable disease.... In thirty-two hours the first [oldest boy] had the affusion fourteen times; eight times cold, twice cool, and four times tepid. Twelve affusions sufficed in the case of the youngest, of which seven were cold. The fever was in both completely subdued. On the morning of the third day they were both evidently safe."

On August 31, 1805, just prior to the fourth publication of his pamphlet, his research came to an abrupt halt. He died as a result of a heart disorder. Yet his cold-water treatment approach continued to spread. It was recommended by physicians for the greater part of the next three decades.

Still, a considerable intuitive component surrounded when and how much to cool patients, and as time went on, physicians became increasingly unable to successfully practice Currie's technique. He was no longer around to advance his methods, nail down an exact procedure, or correct misguided physicians, who may have been practicing based on vague word-of-mouth learning. And so, by the 1830s, Currie's influence began to wane.

More broadly, and in twenty-twenty hindsight, the technique likely fell by the wayside because it simply didn't work. According to current medical knowledge, cold-water bathing is certainly not recommended for fever treatment.

Widespread interest in cold-water therapy did, however, succeed Currie's frontier legacy. And, again, it originated largely outside the medical community. This resurgence can be credited to a peasant farmer in Austria by the name of Vincenz Priessnitz. His cold-water treatments drew much attention, both positive *and* negative.

The story goes that as a boy growing up on a farm in Gräfenberg, located in Austrian Silesia, Priessnitz observed a wire-worker using cold water to assuage a sore, injured hand throughout the day. Similarly, he observed wild animals around the farm submerging injured limbs in the cold waters of a nearby river, for pain relief. Following his own boyhood mishaps, one involving a sprained wrist and another a crushed thumb, he recalled these sightings and repeatedly immersed his own injuries in cold water to discover significant pain reduction.

At the age of sixteen, in an accident involving a frightened horse, Priessnitz was hit by a cart wheel and left with three broken ribs. After regaining consciousness he was told by a local surgeon that he'd never be able to work again. In defiance of this grim prognosis, he set out to treat himself. He replaced his surgical bandages with cold, wet ones, which he replaced again, repeating the process at regular intervals. Painfully, he created a therapeutic routine that required pressing his abdomen against a table while taking deep breaths. He continued the cold affusions, drank plenty of water, and persisted with his pain-inducing rehabilitative exercise, also getting plenty of rest in the fresh country air. He was able to move about with relative ease after just ten days and was able to continue farm life after a year.

His recovery attracted much attention, and it wasn't long before Priessnitz was consulting with others on his cold-water, health-conscious methods. Over the next few years, he developed his practice, and the buzz continued to grow. After a decade, scores of patients were arriving from throughout Europe to reap the benefits of his cold-water cure, as it became known. He used the treatment on patients with a variety of illnesses—from broken limbs to paralysis to insanity. Often the regimen involved being wrapped in wet bandages and blankets to promote sweating, followed by immersion in cold water.

By the mid-1830s Priessnitz had established a spa, equipped with pools and showering devices, some constructed for use on a particular body part, such as the head or the foot. There was even an eye bath. As his fame increased and he became a household name, the spa was visited by thousands of patients, including a steady influx of European royalty. Yet it was also accessible to those with no aristocratic background.

Regardless of a visitor's wealth, the definitions of "patient" and "guest," "therapy" and "treatment" seem to have been obscured between those genuinely suffering from illness and those enjoying the benefits of a spa vacation. After all, Priessnitz had no formal medical background, and his treatment regimes were medically unsubstantiated.

This context drew interest from those skeptical of medical professionals and their practices and jargon. Indeed, Priessnitz is today regarded as an early purveyor of alternative medicine in the form of naturopathy. Eventually, he was tried in court on multiple occasions for being a medically uneducated con artist, but he was acquitted each time. In the end, his acquittals only fueled more positive public attention.

The skeptics argued that he never wrote a single word regarding the physical process of his treatments, let alone put forth any testable, substantive theory about how and why they might work. Rather, he continued his naturopathic attitude, promoting the body's natural healing tendencies and adding some vague notion about cold water ridding the body of toxins.

This basically antiscientific attitude was advanced by Richard Claridge, who wrote a book titled *Hydropathy; or, the Cold Water Cure, as Practised by Vincent Priessnitz* . . . (1842 and 1843). It was as if he were writing a book *for* Priessnitz, in response to the skeptics. Claridge, who had been a patient at Priessnitz's spa, carefully documented both the treatment of others and his own successful visit. Here's a portion:

> That Mr. Priessnitz has founded some sort of theory on his mode of treatment, after so many years of successful practice, and with the help of that inquiring genius, and that natural impenetrable calmness which so particularly distinguishes him, there can be little doubt; and this theory has never failed him in his treatment of the most complicated diseases. But he has no time for writing; and if he had, he would find it extremely difficult to explain himself; since it is an extraordinary fact that no two cases are treated exactly alike. There is no doubt that Mr. Priessnitz owes all his experience to his utter ignorance of medical science, which, indeed, is his greatest advantage; for what does the history of medicine offer, but the discouraging picture of the instability of principles, and a series of theories succeeding each other, without any one of them being able to content an upright spirit, or satisfy an inquiring mind? We can hardly expect, however, that Mr. Priessnitz will ever attempt to give the world any medical or systematic details. This is only left to intelligent persons.

For a book that is still widely circulating and deemed a "classic" (by Amazon, at least), the dubious nature of such fanatical-sounding appeal is notable since it flies in the face of any true scientific value.

Regardless of the wavering foundation upon which the water cure rested, it *felt* therapeutic for many people. Who wouldn't feel better after a spa vacation? The practice soon established itself into what is now known as hydrotherapy: broadly defined as using water—be it cold, tepid, or hot—for the treatment of various health conditions.

Priessnitz had one follower in particular who, still without any scientific underpinning, expanded his hydrotherapy practices into a holistic approach with the aim of better overall health, both physical and mental. His name was Sebastian Kneipp. Kneipp's practices gained an even larger notoriety than the strict bathing and soaking routines of Priessnitz, and he went on to leave a greater legacy—one that exists still today.

As a young man, Kneipp contracted tuberculosis. In a manner similar to that adopted by Priessnitz, he decided he would not settle for the grim prognosis foisted on him by his doctor. After a nearly complete recovery, he claimed he owed the success to the water cure. And so, after being ordained as a priest in 1852, Kneipp continued to explore hydrotherapy as a means of treatment for various ailments. He had some success with patients when he tried a new, milder approach with less extreme temperature shifts than those used by Priessnitz. He also relaxed Priessnitz's strict regimens—it was like a water cure lite. The hydrotherapy, however, was only one component of his holistic approach to treatment and prevention, which additionally involved herbal medicines, exercise, a healthy diet, and mental wellness. Together, these elements comprised the "five pillars" of health, according to Kneipp.

Although many of his water treatments did not necessarily involve hypothermia-inducing temperatures, a central element to his practice required gushes of cold water to specific body parts. He continually refined his methods and in 1886 published a manuscript titled *My Water Cure*, which detailed the approach.

Judging by historical accounts, Kneipp appears to have been considerably more humble and humanitarian than Priessnitz, as evidenced by a more concerted effort toward helping the poor. That said, his clients also came to include Archduke Franz Ferdinand of Austria; his father, Archduke Karl Ludwig; and Pope Leo XIII. Kneipp always acknowledged that he had no formal medical training, and he admitted that his water cure was not in line with current medical theory and practices. He believed that his pillars of treatment and well-being were both preventive and therapeutic and that they "began where conventional medical knowledge ended."

Central to Kneipp's approach was the idea that blood circulation was an essential component of healthy living. He thought disease and illness were largely caused by impurities and foreign matter in the blood, and that cold-water treatments would aid in their elimination while improving circulation and vitality. Primarily, however, his cold-water practices were heralded as a simple activity that aided general health and psychological well-being.

Although today Kneipp's practices are widely considered a part of alternative medicine, which lies outside the evidence-based, scientific establishment, his holistic approach lives on. Known as Kneipp Treatment or Kneipp Therapy, it was awarded a UNESCO cultural heritage status in 2016, and the techniques it promotes are commonly offered at spas around the globe. They are seen less as curative than as therapeutic, in the manner typically associated with spa treatments in general—or with something as nebulous as "fresh air," for that matter.

## COLD-WATER SWIMMING

Cold-water therapy today is moving beyond this traditional practice and undergoing a new wave of popularity.

Over the past decade there's been a steady increase in the chilly practice of cold-water swimming, involving adventurous bathers both young and old. Further, a slew of new studies have shown various benefits of exposure to cold water. More and more people are taking to frigid waters for a variety of therapeutic goals, ranging from the treatment of arthritis to managing depression. What uniquely marks this recent trend is that such benefits are relatively well-grounded; that is, although we may not have a completely detailed understanding of how regular plunges into frigid waters can benefit us, we are beginning to see testable explanations emerge from repeating patterns of data.

Some theorists speculate that underlying the therapeutic value of cold-water swimming is a dose-response phenomenon known as *hormesis*. The idea is that at low doses, certain otherwise toxic or harmful agents can actually have a beneficial effect. They activate mild stress responses that ultimately strengthen biological systems and promote resiliency. Take spicy foods as an example (which lie at the other end of the hot-cold spectrum). They can indeed be harmful in high doses; just imagine the consequences of swallowing a whole liter of pure habanero juice. Because most mammals react negatively to the ingestion of spicy substances, some researchers theorize that capsaicin, the active ingredient that makes a substance peppery, serves as a defense mechanism for plants. At low doses, however, capsaicin has been shown to afford a number of benefits. Moreover, spicy foods are typically consumed in temperate climates in part because although they initially increase the sensation of heat, ultimately they yield apparent cooling effects that begin after the "hotness" disappears.

Although it seems a mystery why many of us masochistically indulge in this pain-inducing compound, we know that at relatively low doses, capsaicin triggers increases in heart rate, breathing, and adrenaline. It gives a jolt that excites, widens the eyes, gets the blood flowing, and engages our inner daredevil. Numerous studies have found that it offers health benefits, including analgesic, antiobesity, antipruritic (anti-itching), anti-inflammatory, antiapoptotic (protecting against cell death), anticancer, antioxidant, and neuroprotective functions. Capsaicin has also been cited as beneficial in treating vascular-related diseases and metabolic syndrome and in providing gastroprotective effects.

High and low doses of cold mirror this pattern in key ways. *Cold shock*, which occurs as one enters frigid water up to their head, is increasingly dangerous depending on the temperature of the water and the speed at which it is entered. As you might guess, the faster the entry, the more shocking, which is why caution needs to be exercised; cold shock can trigger gasping, which can cause one to inhale up to two liters of water, suffocate, and drown. If you don't die that way, the shock causes stress responses including an elevated heart rate, vasoconstriction (constriction of blood vessels), and increases in blood pressure, respiratory rate, and stress-hormone release. Cold shock and the resultant gasping have been put forth as arguably one of the main causes of death in countless shipwrecks, including that of the *Titanic*.

Although cold shock can lead to death, in far less extreme circumstances it can also, in a sense, breathe life by providing known physiological, psychological, and emotional benefits. Recent studies show that mild cold-shock responses, experienced on a regular basis, go beyond their momentary invigorating effect. Such a response to repeated cold-water bathing is known to reduce bodily levels of uric acid, which ultimately can lead to improved mechanisms for coping with stress on a long-term basis. Sitting in cold water after the initial shock period can cause decreases in heart rate and respiration frequency, induce calming, and lower blood pressure—changes that accompany reduced risk for stroke and heart attack. Further, a wide body of literature supports cold-water therapy for pain relief in athletes and in sufferers of acute and chronic pain. A sudden drop in temperature after a workout causes blood vessels to constrict, changing the way blood and other fluids circulate and interact throughout the body. Ultimately, this cascade of reactions flushes away metabolic waste and helps recovery.

In relation to pain relief, cold swimming or cold showering can trigger anti-inflammatory responses that offer a surprising benefit. Inflammation has been strongly linked with depression. Although at first glance it may seem an unlikely relationship, studies show that in response to stress, low socioeconomic status, or a troubled childhood, for example, the immune system initiates a nonlocalized inflammatory response. Instead of a localized inflammatory response that results in, say, a red, swollen appendage, a prolonged nonlocalized response can result in symptoms of depression. New evidence shows that the body's anti-inflammatory responses to cold shock may afford a form of treatment.

## COOLING AND PSEUDOPSYCHOLOGY

The route to improved physical and mental health via cold water, however, hasn't exactly been an idyllic trip down a cool, lazy river on a hot summer day. Beyond the controversy stirred by those who condemned hydrotherapy for its lack of medical and scientific foundation in the era of Priessnitz and Kneipp, its practice came to include an undoubtedly tragic vein.

Remember John Floyer and his tour of countryside bathing spots? We encounter him again at the very frontier of hydrotherapy. His writings about cold water contained some pretty gruesome ideas. Considering the title of his book on the topic, *An Enquiry into the Right Use and Abuses of the Hot, Cold and Temperate Baths in England*, what he counted as "right use" and as "abuse" seem quite subjective. Take for example this passage:

> Amongst many wise men; they shave their heads every week and wash it every morning with cold water, which hardens the skin, and cools the brain whereby the flux of too much water into it is prevented, and that coldness of the head renders it fitter for all rational thoughts, and the animal spirits being compelled are more lively, springy, fitter for motion.

Every parent wishes his child may be bred up to a great degree of hardness. The best methods to attain that, is the immersion, at first, into cold water in Baptism; and afterwards, to use the method of washing their children in cold water every morning and night, till their children are three quarters old, for by this the Welsh women use to prevent the rickets in their children; tis a common saying amongst their nurses, that no child has the rickets unless he has a dirty slut for a nurse.

Besides drawing some rather misguided conclusions about nurses based on children with rickets, Floyer was an advocate of regular cold-water dunkings using methods that were likely to cause more harm than benefit. But just how risky are his implications? Unsuspecting babies aside, those potentially maltreated by the suggestions set forth in the above passage include people with psychological disorders—those whose heads are *not* deemed "fit for rational thoughts." For around two hundred years, beginning in the early 1600s, cold water emerged, evolved, and thrived as a cure for mental disorder. Unfortunately, this "cure" caused significant suffering and even death in its patients.

Going beyond the suggestions of Floyer, cold water was unambiguously advocated by Flemish physician and chemist Jean Baptiste Van Helmont (1580–1644) as a remedy for mental disorder. Van Helmont is perhaps best known for his advocacy of *spontaneous generation*, the theory that life could be created from nonliving matter. He believed, for example, that it was possible to produce mice by combining wheat and a wet shirt in an enclosed vessel. His medical tome, *Ortus Medicinae*, published in 1643, contains his unique and equally senseless conception of the cold-water cure.

In it, he describes immersion in cold water as a treatment for those suffering from "madness." And not just a quick dip, like that intended by Floyer for babies. Van Helmont was inspired upon hearing about a chained "lunatic" being transported by wagon who freed himself and apparently fell into a frigid lake, where he lost consciousness while struggling for his life. After being rescued and revived, he recovered his sanity and never experienced a "further attack of madness." And so, Van Helmont's idea of cold-water therapy was to fully immerse a patient in frigid water until they lost consciousness. At that particular point, when the "patient" (read *victim*) was on the fringe of drowning to death, it was possible to "kill the mad idea," according to Van Helmont. His son's biographical writings suggest that he believed that cold-water immersion in this gruesome manner could stop "the too violent and exorbitant Operation of the fiery Life." Here, one can assume that "fiery life" was his manner of describing the active, erratic behavior associated with psychological disorder.

Van Helmont began a routine of stripping his patients of their clothing, binding their hands, and forcefully lowering them into cold lakes or baths. He believed "the only care that must be taken is to plunge the sufferers into the water suddenly and unawares, and to keep them there for a long time. One need have no fear for their lives." As you can imagine, however, this practice, which was often carried out in public, could kill more than the mad idea. Unfortunately, death was seen only as a potential "side effect."

As the regimen grew in popularity—in England, for example—its practitioners rationalized that cold-water treatment could be performed at "lunatic asylums," which were new at the time, to the benefit of patients and, on a larger scale, the entire English nation. The idea was that curing patients and releasing them expended fewer tax dollars than housing and feeding them for the long term.

The cause of "mania," according to physicians, was an overheated brain, and they believed that cold water could treat it. According to some texts, "overheating" could be brought on by inappropriate thinking patterns, like being overly passionate. William Rowley, a London-born physician, wrote in 1793, "Madness is commonly caused by a gradual, slow, and continued determination of the blood to the brain from pathemata animi, or meditation on one subject, until the fluids of the brain become, and continue, vitiated." And so the solution seemed clear: cool the brain. But how?

During cold water's rise to prominence for these purposes, just about every torture contraption for soaking, spraying, and dunking patients that you could possibly imagine came to be. In remarkable feats of engineering, arrangements with wooden frames, pulleys, and ropes went from design to reality. Victims were bound and immersed, often multiple times. Originally, devices were created to forcibly plunge them into an existing body of water—say, a lake or river—where the treatments were viewed in public by gawking onlookers. As the frequency of such practices grew, dunking stations were built on-site, at asylums. These in-house contraptions concealed any screaming or morbid side effects (i.e., death) from public knowledge.

Water was often poured from reservoirs, installed on rooftops, onto the heads and bodies of patients who were restrained in some manner, preventing them from escape. Some were tied to a chair, execution style, while cold water was flooded on them from above and then drained through the floor. Other approaches involved confining the patient to a container from which only their head protruded—a more efficient design because the water in the reservoir drained at a slower rate, enabling a longer "treatment" session. Because the overheated blood vessels of the brain were thought to be the source of madness, the head also served as a more efficient target for the "fall of water"—no need to plunge the entire body.

The contraptions in their various forms became known as *douches*. One doctor, Sir Alexander Morison, described them in 1828:

The different parts of an apparatus for giving the Douche, con-sisting of a bucket, from which a stream of water is made to fall on the head of the patient from different heights, regulated by a rope and pully,—by the cock inserted into the lower part of the bucket, the size of the stream is regulated.

The douche of cold water . . . directed upon the head, as well to diminish vascular activation in the brain as to repress violence, to overcome obstinacy, and to rouse the patient when indolence or stupor prevails.

Morison, in the final sentence of the quote, runs through a va-riety of psychological afflictions deemed treatable by cold-water therapy. Mostly, however, it was aimed at behavior characterized by an overabundance of passion, will, and obsession—the "mad idea" that was thought to cause "violence in individuals that were healthy and strong." They were seen as powerful and as not need-ing basic resources like food or warmth. In this case, cold water was used to induce fear and break their will. Only then, when subdued—and nearly dead, for that matter—could they be guided toward sanity and reason.

Once a session began, these unfortunate patients were often made to suffer the fall of water for as long as it was thought they could endure it.

Philippe Pinel, a French physician known for pioneering the humane treatment of patients with mental illness, weighed in on the use of the douche in 1794. Not in a manner, however, that one would hope for—his writing on "moral treatment" proposed that "one of the major principles of the psychologic management of the insane is to break their will in a skilfully timed manner, with-out causing wounds or imposing hard labour. Rather, a formida-ble show of terror should convince them that they are not free to pursue their impetuous wilfulness and that their only choice is to submit."

*Moral* treatment?

More broadly, it seems that the douche was primarily used for patients experiencing deep-rooted delusions and hallucinations that resulted in erratic behaviors that put both themselves and asylum workers at risk of a violent outburst. For example, it might be used on people who had paranoid schizophrenia, who experience fear and anxiety resulting from delusional but strongly held convictions that anyone, be they strangers, family members, or even the government, could be plotting against them. Such anxiety may become so intense and prolonged that it can provoke aggressive behavior. But if patients' delusional aggression against an opposing force was any indication of the fear caused by irrational thoughts, sadly, it also might have been an indication of the degree of water torture they were made to endure as a "treatment."

Obviously, not all patients were aggressive, and many who weren't also had to endure the douche. In some cases, it seems there were no clear indications of insanity when the method was used. Take, for example, the following description, written in 1725 by a physician named Patrick Blair, regarding a woman who was not being a "dutiful wife": she was failing the expectations of her husband. Specifically, she was judged as "neglectful." Blair describes how she was "treated" by the pouring of cold water from a basin positioned atop a tower, while she remained bound in a tub thirty-five feet below:

> A married woman . . . became mad, neglected every thing, would not own her husband nor any of the family. . . . I ordered her to be blindfolded. She was lifted up by force, plac'd in and fixt to the Chair in the bathing Tub. All this put her in an unexpressable terrour especially when the water was let down. I kept her under the fall 30 minutes, stopping the pipe now and then and enquiring whether she would take to her husband butt she still obstinately deny'd till at last being much fatigu'd with the pressure of the water she promised she would do what I desired on which I desisted, let her go to bed. . . . A

week after I gave her another Tryal but adding a smaller pipe so that when the one let the water fall on top of her head the other squirted it in her face or any other part of her head neck or breast I thought proper. Being still very strong I gave her 60 minutes this time when she still kept so obstinate that she would not promise to take her husband until her spirits being allmost dissipated she promised to Love him as before. . . . I gave her the 3d Tryal of the fall and continued her 90 minutes under it. . . . I threatned her with the fourth Tryal, took her out of bed, had her stript, blindfolded and ready to be put in the Chair, when she being terrify'd with what she was to undergo she kneeld submissively that I would spare her and she would become a Loving obedient and dutifull Wife for ever thereafter. I granted her request provided she would go to bed that night with her husband, which she did with great cheerfulness. . . . It appear'd that in 90 minutes there was 15 Ton of water let fall upon her.

Although the unfortunate woman may not have been psychologically disordered, her "neglectful behaviour" was replaced by fear, driven into her by the douche, after what must have seemed like endless sessions of torture, pain, and suffering.

Cases like this hinted that the douche had utility beyond the treatment of the insane: for supposedly correcting behavior. Although the woman in the above account was deemed "mad," it could be argued that such a label was placed on her simply because her actions were not in line with social norms or ideals. Similarly, today, behaviors that fail to comport with standard protocols are often considered adverse, harmful, or even criminal. Indeed, the difference between those who commit crimes because of psychological issues and those who are simply criminal has always been subjective. As time passed, cold-water douches came to be used in prisons as well as asylums. Reports of the use of cold-water torture in American prisons tell its long history, with instances recorded until at least a 2007 report by the International Red Cross on the treatment of fourteen detainees in CIA custody that year.

Thankfully, however, by the beginning of the 1900s, cold-water douching came to be seen by many for what it was—torture—and it was used less frequently. Nonetheless, around that time, the practice of simply threatening prisoners with death by water douche seemed to be effective and to some extent overtook actual douching. What progress!

Such changes curtailed the practice of douching and are likely to have saved lives. The various douche contraptions and their component straitjackets, leather straps, and buckles were put away and deemed outdated. Patients were now treated in "hospitals" rather than simply housed like animals in "lunatic asylums."

Although cold-water torture in the asylums was infrequent by the 1900s, the pseudoscientific rationale behind its effectiveness to some extent only grew more elaborate. Around 1915, Eva Reid, a physician in San Francisco, wrote about the "science" behind cold's effectiveness. According to her, it came from "elimination by the skin, helping to rid the system of toxins and poisonous matter in the constitution. . . . [It] brings blood to the surface and relieves the congestion in the brain and spinal cord, which in most cases seems to cause the excitement." What brings such groundless contention to a level beyond the simple eighteenth-century notion of an overheated brain is that in her writing these claims were backed up by physical measurements of heart rate, respiration, blood pressure, and other indices, adding further unsupported reasoning to substantiate the already-false claims.

Indeed, hydrotherapy had achieved a complete, comprehensive, and complex state of pseudoscientific presence; it was an entire scheme, from investigation to application.

It worked because psychological instability was defined largely by a lack of functionality within one's society. Thus, the douche was ultimately seen as a service. According to Pinel's ideals, it was imperative to remind the victim that

> we profit from the circumstance of the bath [shower-bath], remind him of the transgression, or of the omission of an important duty, and with the aid of a faucet suddenly release a shower of cold water upon his head, which often disconcerts that madman or drives out a predominant idea by a strong and unexpected impression; if the idea persists, the shower is repeated, but care is taken to avoid the hard tone and the shocking terms that would cause rebellion: On the contrary, the madman is made to understand that it is for his sake and reluctantly that we resort to such violent measures; sometimes we add a joke, taking care not to go too far with it.

By the 1940s, however, cold-water treatment was rare. New forms of therapy were coming to the fore, such as electroconvulsive therapy and pharmaceutical intervention. And although progress has certainly been made with these approaches, they, too, have run into controversy. Early forms of electroconvulsive therapy—then known as "shock therapy"—for example, caused injuries as patients convulsed wildly. And even today, the exact mechanisms and benefits of various psychoactive medications are hotly debated.

Throughout the centuries, although cold treatments progressed from local applications to complex whole-body arrangements, they were carried out with scant regard toward securing a fundamental, comprehensive scientific understanding of hypothermia. In hindsight, the treatments basically amounted to torture, used to subdue anyone who didn't dutifully conform to the social order. The unscientific practices regarding cold were complemented by the popularization of the "cold-water cure" as a form of treatment offered by an increasing number of spas and often sought by wealthy vacationers.

Still, all was not lost. Looking back on the seventeenth, eighteenth, and nineteenth centuries, crucial developments in science enabled a foundation for therapeutic hypothermia. Chief among them was the thermometer. Though it took a while for science to catch up to the thermometer's full potential, its invention was crucial for future investigations. Going beyond hand thermometry, the use of a calibrated device with a standard scale enabled a quantitative, precise, and accurate measurement of human temperature that could conclusively determine if a person was indeed hypothermic. These developments set the stage for new forms of cold therapy that were more refined and humane, ultimately leading to the scientific foundation of today's means of treatment.

◇◇◇◇◇◇

*Cold has enveloped your entire sense of being.*

*You know that it's imperative to keep moving, even though your hands and feet feel like sore lumps of frozen flesh. Your motions grow clumsy and awkward. But you know you need to stay active or your fate will be sealed in ice—perhaps literally.*

*But first, you need to rest—again—just for a moment. One glorious moment. Maybe lie down for a bit and hope that if anyone is out here, they will find you.*

*You fail to notice the stark lack of sense in your inconsistent thoughts.*

*Then, during a habitual, mindless scan of your surroundings, you see something emerge from what seems like a complete abyss.*

*Lights!*

*They are blooming in the distance, on the horizon. There must be life here after all! As you move toward them, you track them carefully so as not to lose sight. They are blurred and unfamiliar, unlike lights from any object that you can recognize. What on Earth could they be?*

*You orient your progress toward them as they beckon you like the visual equivalent of a mysterious siren. But the lights go out. You stare in denial until your eyes begin to water.*

*The desire to rest again increases exponentially. Surely a quick break will provide invaluable respite. Afterward you'll feel revitalized and renewed. You'll be so fast and steady on your feet that you'll end up ahead of where you would have been without having rested.*

*No? Surely the lights will reappear. You must be so close now.*

# 3

# "TREATMENT" AT 90°F (32°C)

## THERAPEUTIC HYPOTHERMIA EVOLVES FROM AN ART TO A VOLATILE SCIENCE

By the early twentieth century, the means for acquiring an accurate, detailed, and, most of all, reality-based understanding of hypothermia and its potential as a lifesaving therapeutic tool were slowly falling into place. The scientific method, which we now understand simply as *science*, was becoming ubiquitously evidence based, quantitative, empirical, and theory driven. With the invention of new technologies like refrigeration, and with innovations in the emerging field of microbiology, the physical effects of hypothermia on human tissues could be understood more fundamentally than ever. Exploration of the effects of cold on basic cellular processes began to provide the physiological groundwork needed to unleash its therapeutic potential. The missteps of past experimentation helped to refine approaches that guided cold therapy along the lines of irrefutable facts. In some sense, over the early twentieth century, therapeutic hypothermia came of age. But with big discovery also comes big responsibility. Such responsibility was not always heeded.

Incredibly, despite the remarkable progress, the twentieth century would also witness more episodes of cold used as torture—as well as more controversy surrounding cooling—than ever before. In a dramatic series of events involving staggering amounts of both advancement and regress, approval and disapproval, acclaim and stigma, induced hypothermia would both cure and kill countless humans and nonhuman animals.

## TEMPLE FAY: CRUCIAL ADVANCEMENTS, ANGRY NURSES

In 1919, an eager sophomore in medical school was posed a question during a verbal quiz. He had no idea that this simple event would impact him for the rest of his life. It would ultimately set the course of his future, and that of many others who followed.

The question concerned cancer: Why is metastatic cancer rarely found in the limbs beyond the knees and elbows? After careful reflection, rather than trying to bluff an answer, the student admitted that he didn't have a clue. Perhaps in appreciation of his honesty, the professor admitted that he, too, did not have any real idea. But that was not the end of it. The question simmered constantly in the student's mind.

The young academic was Temple Fay, born in 1895 in Seattle, Washington, and attending the University of Pennsylvania Medical School. His lineage, traceable back to the 1600s, was well stocked with academics, scientists, and naturalists. Surely his eagerness was born at least partly out of expectation.

And he delivered. Fay became a professor of neurology in 1929. A decade had gone by with that question in the back of his head.

It wouldn't be until the latter half of the next decade, however, that he had a eureka moment. Fay realized that the inability of tumors to form and grow toward the end of one's limbs might have something to do with temperature. Specifically, the temperature of one's hands and feet being usually lower than one's core body

70

temperature. Following this insight, Fay devoted much of his time to researching the temperature of both cancerous and healthy tissues from various locations on the body, and indeed he showed empirically that lower limbs, which often had lower temperatures relative to one's body core, were less likely to be cancerous.

But this discovery wouldn't be the end of his cancer research; his findings brewed with potential. Fay took the obvious course of action: he conducted physiological research at his farm in Maryland (okay, maybe not the *most* obvious course). There, he created a laboratory to investigate cellular activity. He monitored and examined the growth of chicken embryos at various temperatures, finding that below 90°F (32°C), cellular differentiation—the ability to grow increasingly complex cells with different functions—ceased nearly entirely. Then, using tissue culture methods, he found that cooling consistently inhibited the growth of tumor cells. It was a breakthrough!

Now he needed to take his research from the lab to the clinic. But *how*, exactly, to implement a cooling regime to treat cancer? Although Fay's excitement was genuine and motivated by sound reasoning, the task brought with it both technical and practical challenges that must have seemed overwhelming. Although refrigerators and archaic forms of air conditioning were available at the time, artificial cooling technology that would be suitable for this job—that of cooling the bodies or body parts of living cancer patients—simply did not exist.

Yet Fay pressed on. He aimed to treat patients suffering from terminal cancers that were deemed inoperable. The idea was to surgically insert devices into the body that would induce cooling of the cancerous tumor with the hope of stopping its growth. Fay also surmised that cooling would reduce inflammation and possibly contribute to a reduction in pain.

His inaugural cooling device was quite a hack. It consisted of an emptied $CO_2$ capsule—like those used for air guns. Two metal openings were soldered to one end and connected by rubber tubing to a water cooler. Ice and water were placed in the water cooler and made to circulate by gravity through the capsule via the tubing. The capsule, surgically inserted into the body, applied the cooling effect directly to the tumor. Fay called his devices "bombs." Developed by absolute do-it-yourself methodology, they obscured the border between genius and jury-rigging.

The first patient to be treated by one of Fay's bombs was a woman who had cervical cancer, which exhibited as a large, tumorous pelvic extension. The device was put in place, ice water was continuously pumped through it, and the patient was monitored regularly. Incredibly, after forty-eight hours she reported herself as pain-free. After five days, the tumor had apparently shrunk. The procedure was considered a success, and Fay was motivated to treat other patients.

Using what at the time were advanced surgical methods, he set out to insert hypothermic cooling bombs into cancerous areas that often lay deep within patients' bodies. He would, for example, surgically implant a sterilized bomb within a patient's skull alongside their brain via craniotomy (opening of the skull). He developed other bombs of various shapes and sizes, suited for specific procedures, and experimented with different refrigerants and pumping methods. In some patients, the bombs were left in place for weeks and maintained at a chilly 45°F (7°C). Fay's implements were used to treat abscess, cerebritis, osteomyelitis, and, of course, cancer. They reduced pain and stunted tumor growth, and when implanted in the head they were reported to be well tolerated by patients' brains, which generally didn't show signs of rejection such as infection and swelling.

Some tumors, however, lay so deep that they were simply inaccessible and could not be treated by these localized methods. Something more needed to be done. Fay's solution was astounding. He developed whole-body refrigeration regimes for these patients—yet again, without any established technical means. He hypothesized that for cooling to be effective, the patient's core temperature needed to be no higher than about the mid-80s Fahrenheit (lower 30s Celsius). One of the many problems he faced was that clinical thermometers made for measuring body temperature didn't go that low. They were calibrated to 95°F (35°C), colloquially known as the "thermal barrier," the lowest temperature at which human life was thought to be sustainable. Thankfully, however, Fay was able to create a new thermometer design that bottomed out at a lower temperature; soon the custom instruments were being built solely for his purposes.

His first patient for the whole-body protocol, a woman suffering from widespread metastases, was treated in November 1938. In a closed hospital room with the heat turned off and a window left open, her bedding was placed over a drainage system and she was placed under 150 pounds of ice chips. She received a sedative mixture and then entered what Fay and his colleagues called "frozen sleep." Her core body temperature slowly fell until it reached 90°F (32°C). Signs of life soon became only whispers and hints of basic functioning—her pulse all but disappeared and breathing became minimally observable. After the patient was subjected to eighteen hours of refrigeration, Fay feared that she would suffer brain damage due to a lack of oxygen, and so she was carefully rewarmed. Once she regained consciousness, she had no recollection of the experience and showed no apparent signs of brain damage.

Not only did the patient avoid suffering any ill effects; the procedure appeared to show some success in treating her tumors. Essentially, Fay broke the thermal barrier. This triumph gave him great motivation to continue the experimentation. Fay's crudely hacked ice-bucket approach, however, didn't go over well with the ward's nurses. It was winter, and he had instructed them to keep the windows open to aid the cooling process. In addition, the frozen sleep sessions grew longer, which meant the ice levels had to be maintained, water had to be continuously drained, and the patient had to be monitored in various ways—day and night, with the frigid air blowing in. Before long, virtual mutiny broke out, and the nurses, who reported getting colds and suffering illness themselves, staged protests.

According to Fay's own recollection,

> Frankly, the nurses were scared: they could not get the patient's temperature with the clinical thermometers. "The long-stem laboratory thermometers might break in getting a rectal reading. The ice and ice water were always in the way, even when the patient was turned. The pulse was weak. The breathing was shallow. I can't get the patient's blood pressure," were all reports that kept the staff in a constant state of alarm. The entire project of general refrigeration had snowballed into a vast issue of distortions of truth, and even my friendly colleagues began to look askance and asked how long was this absurd experiment going to be permitted.

Although Fay is said to have acknowledged the determination, effort, and perseverance of his staff, the complaints of the nurses were influential, and the program teetered on the brink of termination until technical improvements made the process more practical. Specially built cold blankets housed tubes through which ice water was pumped; they were kept chilled via a commercially available beer-cooling system adapted for the purpose.

Astonishingly, the new system kept patients at temperatures between around 91°F (33°C) and 77°F (25°C) for up to eight consecutive days.

Stepping back, one wonders: How could such an experience not be deadly, let alone be therapeutic?

Some context reveals the initial patients were actually nurses who had been diagnosed with terminal cancer. With knowledge gained from their experience treating other cancer patients, they knew what they might otherwise be in for, and they understood that the refrigeration treatment wasn't an alternative they could readily dismiss. They knew that great care would be taken to keep them unconscious during the entire refrigeration period, through the administration of doses of sodium bromide. The patients were also evaluated psychologically before the inductions in an attempt to ensure that they were mentally and emotionally fit for the procedure.

Between 1936 and 1940, Fay administered localized and general whole-body refrigeration therapy to 126 patients in 169 treatment sessions. The vast majority—95.3 percent—reported pain relief, and 20–25 percent experienced a slowing of cancerous growth. These numbers indicated a promising effect, and although Fay's experiments originally met with a chilling scorn in the medical community, the therapeutic implications of the data could not be ignored.

As Fay remembered it, "What we learned after breaking the human thermal barrier on the hypothermic side, was that human survival was possible under proper supervision. When total body refrigeration was established above 24 C and that a hypothermic state could be maintained for 10 days (probably longer if required) when temperature levels of 29.4–32.3 C were maintained."

Despite Fay's auspicious initial achievements with pain relief and the slowing of cancerous growths, his research program ground to a halt. The full potential of his cold treatment would never be fully realized or put into practice. In fact, his research would end up stigmatized for decades. How did that happen, you ask? Blame the Nazis. No, really!

## DISASTER STRIKES

In 1939, Fay presented his hard-earned, promising results at the Third International Cancer Congress, where they received considerable attention, both positive and negative. Following the conference, he sent a manuscript detailing his research to a publisher in Belgium, where it was obtained by the Nazis, who were gaining control of the territories neighboring Germany. Belgium would be captured in 1940.

The Nazis' interest in hypothermia research was sparked by *Luftwaffe* pilots engaged in air-to-air combat. Although the pilots were able to eject from their planes after being hit, they often met their doom while stranded in frigid waters as they waited for rescue. What intrigued Nazi officials about Fay's data was that they showed that survival was possible at temperatures below what had been previously thought fatal; the data therefore had useful search-and-rescue implications. They wanted to know how long soldiers could endure immersion in frigid northern seawater in order to gauge the likelihood of successful recovery attempts. This set the precedent for a horrible turn of events in terms of research into therapeutic hypothermia.

In 1942, Nazi scientists initiated an immersion-hypothermia project at the Dachau concentration camp. The project, which lasted into the next year, involved between one hundred and three hundred prisoners, consisting mostly of Jews and captured Russian soldiers. These subjects were forcefully immersed in water with temperatures as low as 36.5°F (2.5°C) and only as high as 54°F (12°C) for extended durations, to test their survival.

The existence of these experiments was discovered by Allied forces only during investigations that took place after the war ended. They were unearthed by Leo Alexander, an American psychiatrist, neurologist, and educator at Harvard University. Alexander served as an army medical investigator during World War II. His inquiries into Nazi crimes were central in conceiving the Nuremberg Code of Ethics, which has revolutionized the treatment of human subjects in experiments.

Before the Nuremberg trials, Alexander was tasked with revealing the gruesome details of exactly what had taken place at Dachau. He interviewed German scientists and physicians, who explained that they had been working on cold-water immersion experiments with animals as test subjects. Alexander asked whether any human subjects had also been used in the experiments. He was immediately met with the response "No."

Alexander learned about a radio broadcast during which freed Dachau prisoners described victims who had been forced to immerse themselves in ice water. This enabled him to gather more intelligence. He verified the claims in documents found in an underground vault in Austria, where they had been hidden by the Nazis toward the end of the war.

The hypothermia experiments were described in detailed correspondence between a Dr. Sigmund Rascher, the project leader, and Reichsführer Himmler, commander of the Schutzstaffel (SS) and superior of concentration camps. Rascher, an SS doctor and *Luftwaffe* member, had a unique connection with Himmler through which he gained privilege and influence. It is reported that Rascher's wife, Nini, was Himmler's former lover. Indeed, throughout his career with the SS, Rascher had no inhibitions regarding personal requests in his letters to Himmler. They included everything from a new apartment to fresh fruit and even a servant.

The correspondence contained a fifty-six-page "scientific" report about the Dachau hypothermia experiments. It revealed that in 360–400 experimental sessions, up to three hundred prisoners had been subjected to testing in frigid waters, implying some had been immersed multiple times. Some subjects were dressed in pilot suits and given flotation devices, whereas others wore nothing. Some experiments were conducted in open-air settings within the camp: the prisoners were stripped naked and confined to freezing conditions in outdoor pens for up to fourteen hours. Records of core body temperatures were as low as 79°F (26°C). In one letter, Rascher suggested to Himmler that it "could be wise to move the experimental program to the camp at Auschwitz," because "that camp is bigger, with more secluded spaces so the screams and moans of the subjects are less likely to arouse attention."

In 1946, after the discovery of these and other gruesome experiments at the concentration camps, the Nuremberg trials commenced, and the guilty were sentenced to prison or executed.

What became of Sigmund Rascher? After falling out of favor with Himmler in the final days of the war, he was, in an ironic twist of fate, himself imprisoned at the Dachau camp. Shortly thereafter, Himmler ordered his death. Rascher was immediately shot, days before the end of the war.

Incredibly, the data from the Nazis' hypothermia experiments, as unethical as they were, lived on for many years and were cited in subsequent scientific studies. According to one review, by 1984 the Dachau data had been cited in at least forty-five articles. Some authors believed that the Nazi research provided valuable insight that could ultimately be used to save the lives of hypothermic, cold-water drowning victims. At the same time, one could not ignore its extremely sordid, immoral context, and the implied belief that use of the data only rationalized the means by which they were obtained. The arguments against using the data were as numerous as the ones advocating their potential role in future research. They ranged from detailed philosophical claims invoking Kantian ethics regarding the unjustifiable use of humans as a means to an end, to simple rhetorical statements that appealed to emotion more than to logic and reason.

Consider a statement made by Eva Moses Kor, a Holocaust survivor who, along with her twin sister, was subjected to experiments at Auschwitz: "I am appalled by anyone who seemingly is justifying the means by using the results of the Nazi experiments. In Auschwitz we were treated like a commodity; the hair was used for mattresses . . . the gold collected from the teeth of the dead went into the treasury, and many of us were used as guinea pigs. Today some doctors want to use the only thing left by these victims. They are like vultures waiting for the corpses to cool so they could devour every consumable part."

Indeed, one can imagine how a doctor's opinion about using the data might be affected by the discovery that their own relatives or family members had been victimized in the experiments—that they had met their fate in such a horrible manner by such repugnant hands. Would further use of the data set an evil precedent for advocating unethical practices of the past?

Alternately, the case *for* using the Nazi data was assembled with, arguably, equal conviction. A common philosophical standpoint regards using the data with a utilitarian objective, procuring the "greatest good for the greatest number of people." This reasoning was reinforced by Robert Harnett, of Louisiana Technical University. Harnett, who cited the data in his own hypothermia studies, wrote, "If it's possible that some good can come from it so that lives are saved, then that should be brought about. . . . The other side of the argument is that it condones murder in some way. To me that's just a ludicrous argument. No sane person could condone murder as a means of gathering data. Looking at it from a historical perspective, we can't save the lives of the death-camp victims; they're dead and gone. So the question becomes, 'Can we do any good with it now?'"

Currently, among ethicists, scientists, and even certain Jewish groups, there appears to be some consensus on the matter. In broad terms, if firmly evidence-based, scientific reasoning demonstrates that the Dachau data can be used to significantly advance lifesaving techniques, then doing so can be justified, as long as citation of the data clearly acknowledges the victims, how the data were obtained, and the human cost involved.

## "REFRIGERATION THERAPY"

Leo Alexander was a key contributor to the establishment of the Nuremberg Code of Ethics, which represented a revolutionary advancement in human rights. Among other principles, the code establishes one's right to consent to or refuse participation in any experiment sanctioned by a government, corporation, or academic institution, a right now usually taken for granted. Alexander's contribution toward the innovation of the code was influenced by the Dachau hypothermia project and his discovery of the brutal experiments that had taken place there. Yet it was also shaped by his discovery of other controversial and outright unethical experiments involving cold, ones that occurred within his native country, the United States. Unfortunately, however, the Nuremberg Code held little significance in the practice of human experimentation for the first two decades after its inception, and cold-related injustices continued during that time.

It turns out that not only would Fay's innovative, progressive findings be co-opted by Nazis for their horrifically inhumane experiments; completely aside, they triggered a resurgence in using cold as therapy for mental illness. In a manner reminiscent of the barbaric practices surrounding the douche, the centuries-old, but groundless theory of treating patients with psychosis by chilling them nearly to the point of death reemerged in the twentieth century.

This time, however, with the advent of sedatives and new, improved technology involving automatically thermoregulated cooling devices (e.g., Thermo-Rite blankets), subjecting patients to the cold didn't have to be a merciless process involving the near drowning of screaming victims tied up in leather straps, chains, and buckles. No longer did the protocol yield messy, flooded basins. Most of all, combativeness from patients wasn't a pressing issue. There was no need to "break their will" or even any desire or reason to induce pain or punishment; the motives were, in fact, genuine applications of treatment for those with psychosis.

The difference? Now barbiturate drugs were available that could subdue patients to the point of rendering them completely unaware of a drop in body temperature.

The experiments took place at McLean Hospital, an affiliate of Harvard University, located in Belmont, Massachusetts. McLean, which still exists today, is a comforting, inviting manor—the opposite of a grim, intimidating institutional asylum that one might assume. In fact, by the 1990s, McLean had earned a reputation as

"America's Premiere Mental Hospital." It was known for its roster of celebrity patients, such as James Taylor and Ray Charles. Currently, it houses the world's largest neuroscientific and psychiatric research program in a private hospital and is Harvard's largest psychiatric facility.

In the early 1940s, two Harvard researchers, John Talbott and Kenneth Tillotson, began a cold-therapy program there. Talbott had previously investigated the effects of heat on the psychological state and the effectiveness of men in various work environments. The scientists' interest had shifted to cold; in part, they were motivated by the same sources as those who spawned the douche as a form of "treatment" in the early nineteenth century. Specifically, they were intrigued by an 1805 case in which a psychologically disordered patient was submerged in cold water, a practice that in retrospect they could have dismissed as nothing more than torture.

They became aware of Fay's experiments and his breaking of the thermal barrier and thought cooling was worth a try. Again. Unlike those early, gruesome practices, however, they aimed to use a kinder, gentler approach that employed sedatives and careful, extensive monitoring of vital signs during and after the procedure. They also limited their initiative to patients who had a serious, intractable mental state and for whom alternative practices, such as insulin shock, had failed. In essence, it would be a last-ditch effort. The patients recruited seemed hopelessly delusional, negative, and impulsive. Many had hallucinations, and none were remotely grounded in reality.

Talbott and Tillotson set forth, using innovative rubber blankets they called "mummy bags" because the contraptions enveloped the patient. The bags circulated refrigerants with temperatures between 28.4°F (−2°C) and 23°F (−5°C), making it possible to lower skin temperature (as distinct from core body temperature) by more than 20°C. While cool and unconscious, the patient received a glucose solution to aid in metabolism to prevent cold-related complications. Their temperature was carefully monitored; in later experiments heart and brain activity were also monitored via EKG and EEG technology, impressive for the time. All in all, the technique resembled Temple Fay's approach, but with some advancements. Surely, any untoward complications would be foreseen.

Talbott and Tillotson's first experiment involved ten patients who each underwent one to three sessions within the span of a few months. Nearly all of them were chilled for at least ten hours at body temperatures below 84°F (29°C). Some were cooled for up to thirty-eight hours, and within this group some were cooled for up to seven hours at body temperatures as low as 79°F (26°C). The coolest temperature reported was 73°F (23°C).

The outcome? Talbott and Tillotson reported temporary improvements in nearly all the subjects; four experienced an enduring improvement. In fact, the results were publicized in a 1942 *Scientific American* article that gave their efforts a rave review:

> For the first time this new therapy, popularly known as human hibernation and technically as hypothermia, has apparently found a definite, valuable application—in treating insanity, particularly schizophrenia or dementia praecox. Results in the cases studied have been remarkable, so that more extensive investigation of the possibilities and limitations of this treatment will surely be forthcoming at the war's end, if not before.

Talbott eventually became editor-in-chief of the *Journal of the American Medical Association* and editor of *Psychiatric Services*, and he was elected president of the American Psychiatric Association.

The duo's early success inspired teams at other hospitals to conduct similar experiments—for example, a 1941 study by Drs. Bruce Dill and William Forbes at the Harvard Fatigue Lab.

In a medical-journal article regarding the study, they describe in detail a procedure involving inhospitable environmental temperatures that sounds eerily and disturbingly reminiscent of eighteenth-century victimization:

> While shivering may persist even at 30° [F, or −1°C], the attacks become less frequent and are usually less intense below this temperature. Voluntary movements, sometimes requiring restraint, continue at 30° and even below. Our metabolic measurements probably do not reflect the greatest periods of activity, for then it is impossible to keep the face mask in place. In addition to the voluntary movements and the shivering, it was often difficult to straighten the arm enough to draw blood from the antecubital vein. . . . This state can hardly be one of voluntary contraction, since the performance of so much static work would bring on exhaustion.

Indeed, despite its seemingly rave review of the technique, the previously quoted *Scientific American* article also mentions the level of danger and precariousness involved, but seemingly couches it in profundity: "In human hibernation, life comes as close to death as it is possible to go—and return safely. In the weird world of cold life, you are suspended in the mysterious dark that lies mid way between life and death."

In one case, life came too close to death. In June 1940, a patient succumbed to the treatment. He was a forty-eight-year-old man who was cooled for an astonishing fifty continuous hours before experiencing a circulatory collapse, a fact that seems to have evaded mention in the *Scientific American* article. Nonetheless, Talbott and Tillotson defended the tragedy by arguing that the patient's death was likely caused by circumstantial factors. They declared their investigation a success, saying that it offered a "modicum of hope" for the treatment of patients with severe mental disorder.

In 1943, Douglas Goldman and Maynard Murray—psychiatrists at the University of Cincinnati—conducted a separate refrigeration experiment, again with fatally disastrous consequences. Sixteen patients were cooled in ice and put into refrigerated cabinets for up to forty-eight hours. After their removal, two patients died of complications arising from pneumonia, and several were reported to have sustained permanent brain damage.

With these experiments, the practice of cold was taken too far. It was a bucking bull in the land of cowboy science. The experiments exploited the scientific progress Temple Fay had made and the potential he had brought about for therapeutic forms of hypothermia. His practices, although controversial, extremely progressive, and dangerous, were backed by evidence showing that cooling retarded and abated the growth of tumors and helped to alleviate pain caused by cancer. Through his laborious physiological research with tissue samples, he gathered key insights regarding how and why cold had such effects. The offshoot experiments, although conducted by esteemed researchers, held little foundation in any known reality. They were basically pseudoscience, practiced on patients with no concern for their consent or approval, regardless of other efforts made toward their safety and benefit.

Still, the experiments continued. Chilling patients in extreme conditions held potential as a cure for insanity into the late 1950s. Yet even at that time, the logic on which such practice was based was questionable. A 1956 study conducted at the Central Islip Hospital in New York and authored by Thomas Hoen offered no more explanation than to say, "Just why a state of [cold] narcosis has a tendency to produce improvement in schizophrenic symptomology is not well understood. One interesting theory is that, during the state of narcosis, changes take place in the enzyme system of the central nervous system, tending in some obscure way to reverse the schizophrenic process."

A 1962 review paper by Hungarian physician Peter Veghelyi on "artificial hibernation" shows that not only was there insufficient knowledge regarding a possible mechanism through which cooling could possibly cure insanity; there was also insufficient knowledge about the ultimate fate of such extreme cooling—death. He wrote, "The lower the level of hypothermia under 25 C. [77°F] and the longer it is maintained, the lower is the survival rate. The cause and the mode of death at low temperatures is unknown."

Yet the article's very next statement expresses an extremely dangerous credulity: "A method to ensure revival after protracted hypothermia below 10 C. [50°F] would provide an incredibly effective therapeutic tool."

True, other forms of treatment for psychiatric patients were deemed ineffective, but refrigeration put them at risk of death while also seeming largely unfounded. Still, despite a stark lack of evidence to bridge what must have been regarded even at the time as an enormous logical gap between the physical effects of cold temperatures on human tissue and their application as a treatment for severe psychological disorder, the experiments spanned over two decades. And despite a vast increase in understanding how cold affects the human body, the grounds supporting cold's effectiveness for treating mental illness did not appear significantly more solid than they had in the seventeenth century. Having never succeeded as a treatment, cold therapy for the improvement of psychological well-being was disregarded in the wake of advancements in pharmaceutical approaches, which proved more effective.

In light of the newly established Nuremberg Code, couldn't such a deadly resurgence of experimenting with cold as a form of treatment for mental illness have been prevented?

Perhaps not. Although the code deems that potential subjects must provide *informed consent* before experimenters can engage in any testing on them, it says nothing about their level of understanding or their decision-making capacity. Currently, this means subjects must be informed about the purpose of the experiment, what is required of them, what is involved in the testing, and the potential risks of participating. But even when experimenters inform participants as thoroughly as possible, such an effort can't guarantee a sufficient level of understanding on the part of the subject. At the time of the relevant hypothermia experiments regarding psychological treatment, it was likely assumed that the patients, who mostly suffered from schizophrenia, lacked sound reasoning ability and were incapable of deciding whether or not to consent. Instead, the doctors, who were assumed to have the patients' best interests in mind, decided for them.

The code also stipulates against experimentation that puts subjects at risk of injury or death. In the context of the hypothermia experiments that cost subjects their lives, however, this point is arguably a nonstarter. Ultimately, hypothermia was intended as a form of *treatment*, and careful and comprehensive monitoring of subjects' vital signs during the ostensibly dangerous procedure would supposedly prevent any adverse effects. The fact that such experimentation went on for over two decades indicates that death was regarded as incidental more than it was seen as a major risk.

Nowadays, an established, detailed questioning and assessment process determines whether someone actually carries the decisional capacity required to provide genuinely informed consent to any experiment or procedure. If a patient appears unable to understand the possible risks involved in a particular therapy, or if they're a minor, the decision can be made by a relative or spouse—assuming that person has the necessary decisional capacity.

## EUREKA!

By 1944, numerous medical research programs in the United States were put on hold as a result of the war effort. Temple Fay was forced to slow his efforts toward advancing therapeutic hypothermia. But he remained optimistic, writing, "The wide application of cold therapy almost 100 years ago, when ice was a luxury, reflects today that every human tendency [is] to ignore what is plentiful, common, and easily at hand. The field of refrigeration or hypothermy is broad and deep, awaiting exploration by those who have modern facilities."

The Dachau project and other unethical whole-body human experiments so greatly stigmatized hypothermia research in broader medical circles for decades following the war that it was largely abandoned. But new approaches focused on the cooling of specific, localized parts of the body began during World War II and continued long after. In this way, cooling as a science was loosely returning to its earliest recorded roots in ancient Egypt. This time, however, scientists were equipped with both the knowledge and the technical ability to gain scientific evidence that would seem incomprehensibly advanced to any doctor from that era.

Surprisingly, therapeutic hypothermia's big comeback began with developments in diabetes research. How? Diabetes is characterized by abnormal amounts of blood sugar (glucose). In short, when there is too much glucose in the body, blood flow to the extremities becomes restricted. After time, the tissue can become infected and can begin to progressively rot because of gangrene. Should this happen, the victim's limb usually requires amputation.

Frederick Allen, a noted expert in diabetes research before the widespread use of insulin (the hormone that regulates the body's use of glucose) for treatment, showed that with the aid of a simple tourniquet, chilling a limb in a blood-restricted area rendered nerves and blood vessels resistant to infection because of cold's reductive effect on bacterial cell growth.

Working with Lyman Crossman, a surgeon at a New York City hospital, Allen advanced his cold research by applying it to amputative procedures. The pair showed that refrigeration could reduce the risk of shock during and after the loss of a limb and abate the risk of infection. In addition, cold attenuated nervous system activity, thereby providing an effective analgesic. In fact, amputations undertaken while cold was being applied were carried out while patients remained conscious.

Their innovative methods lowered the mortality rate associated with diabetes-related amputations, often necessitated by severe shock and infection. In their study, 154 patients underwent the technique with only seven deaths—a reduction in the fatality rate from around 65 percent to just 15.5 percent.

Going beyond diabetes patients, these findings were directly relevant for war casualties requiring amputation. The tourniquet technique was fairly quick and simple, and refrigeration technology—or just plain ice—was not difficult to secure on the battlefield. Amputations could be carried out with relative simplicity and efficiency. Alexander R. Griffin, author of *Out of Carnage*, a 1945 book on wartime medical practices, writes of the protocol:

> The ice anesthesia is just about as complete as a pain "blackout" can be. The patient feels nothing, not even when the nerves are severed and the bones sawed. With the nerves in an almost frigid state, the tissues cannot transmit pain impulses. The post-operative condition of the patient does not vary too much from that of others where another type of anesthesia has been used. Healing is somewhat slower than in ordinary operations, but the progress steady. However, the patient's contact with cold still continues. The stump is surrounded with ice to minimize post-operative pain and swellings.

In 1946 Allen and Crossman, concluding their research, reported that cooling reduced cellular metabolism and thus the need for oxygen delivery. This insight could not have been more crucial to the future development of therapeutic hypothermia. It implies that cooling slows adverse effects resulting from a lack of oxygen. By 1954, Allen and Crossman's findings were applied to brain tissue. That's when two surgeons in New York, Hubert Rosomoff and Duncan Holaday, made another key discovery: hypothermia lowers cerebral oxygen consumption.

This realization doesn't obviously spell out anything revolutionary, does it? Yet it set the stage for unleashing hypothermia's protective potential against a lack of oxygen during conditions like cardiac arrest, aneurysm, or stroke, or even when blood flow must be temporarily stopped to conduct heart or brain surgery. And that's only the beginning of what we know now in terms of cold's therapeutic ability. These therapeutic, physiological properties of hypothermia, discovered in rigorous scientific research, are today accepted and routinely sought by surgeons.

Conclusively, twentieth-century advances in science enabled an accurate, elementary understanding of thermal energy. Cold science really took off. It entered a new era that witnessed both great triumphs and deep pitfalls. These discoveries have been crucial in developing various therapeutic applications that are now common and that have saved innumerable lives and prevented incalculable amounts of brain damage. And, as we will see, perhaps even more fundamentally they have blurred the boundary between life and death.

◇◇◇◇◇◇◇

*You're getting colder and colder no matter how much you keep moving. Pressing on only seems to make you more exhausted.*

*The lights now seem to have come from your imagination, even though you recall seeing them as vividly as any reality. They simply must have been there.*

*Absurdly, you can't recall the beginning of this seemingly impossible journey. You don't remember preparing,, when you started, or even whom you were last with. Everything seems blurred. You can't think clearly. It probably doesn't matter.*

*You're transfixed by your vague surroundings. You can't make out any objects, let alone anything familiar, and yet you're mesmerized. Although you keep walking it seems as though the world is moving past you while you passively stare into space, hypnotized. You realize that you are in a stupor. You shake your head to snap out of it. The movement throws you off balance. You stumble. As if in a slapstick comedy, you teeter and fall as though drunk. You feel pain, but even more you feel a looming helplessness, like some external power has control and is deliberately adding to your misery.*

*While you're down, the urge to rest seizes you once more. You start to believe that you're beyond the point of no return. How cold can one become before surrendering to it?*

# 4

# PRESERVATION AT 82°F (28°C)

## HUMAN HIBERNATION AND HYPOTHERMIA FOR PROLONGED SPACEFLIGHT

As I type these words, here at my desk, I'm looking out a window at the houses dotting a valley that reaches toward an expanse of mountains. It's December 20. Since mid-October everything has been covered in a white, fluffy blanket of thick snow. The spruce branches in the forest are quite weighed down, the wind is calm, and the world appears frozen-still, save for some chimneys that slowly puff out smoke.

Although it's just past noon, what makes the houses most visible is the glow coming from inside each window. It's barely light outside.

Here, in the Arctic region of northern Norway, the sun no longer rises at this time of year. It hasn't appeared over the horizon since November 27, and it won't again until January 15. This period, which often includes some of the year's lowest temperatures and harshest storms, is called *polar night*. Locally it is known as *mørketid*, or "dark time," and it couldn't be better for motivating anyone to do . . . absolutely nothing.

Nothing but, perhaps, staying in and sleeping until daylight and warmer temperatures return. It sounds like the perfect context for entering into hibernation—for effectively placing time on hold until it is light enough to resume work, leisure, and, well, everything one might want to do in daylight.

Traditionally, during polar night, times were tough. Any agricultural undertaking was put on an extensive hold, and conditions were often less than ideal for hunting and fishing. Meat and fish from previous seasons had been preserved by smoking and curing, and any fresh fish or meat could be stored in the natural freezer accessed simply by opening one's front door. Thankfully, the restrictions placed by cold and dark on the amount of work that could be done often translated into relatively low nutritional requirements.

It sounds incredibly boring! Many would prefer to simply skip that part of the year. Would it not be ideal to hibernate through it and wake up when conditions are better, when spring returns and life flourishes again? A plethora of animals do it: snails, queen bees, garter snakes, box turtles, bats, and bears (which actually wake from time to time over winter).

Hibernation is a phenomenon typically associated with hypothermic body temperatures that coincides with the onset of colder outside temperatures. Hibernation is normally defined by a significant drop in body temperature during extended bouts of torpor—a prolonged state of lowered physical, mental, and metabolic activity relative to an animal's waking state. Some species, however, don't experience significant cooling during periods of extended rest. Bears, for example, are widely known as hibernators, but scientists have debated whether they are or not because they experience only a slight core temperature drop and wake relatively easily.

The phenomenon of hibernation was known to humans at least as far back as Aristotle's lifetime. Indeed, he wrote of how "animals take their winter sleep . . . by concealing themselves in warm places," noting that brown bears sleep in dens throughout the winter. Because he lacked any other explanation for the disappearance of certain small birds, he incorrectly surmised that those species must also hibernate. Either that or he believed, rather imaginatively, that they transmuted into other species that appeared during the winter. The common poorwill, however, is the only known hibernating bird, and Aristotle was unaware of it. And, of course, no animal can transmute into another species.

Presently, we possess extensive knowledge about which species hibernate. The list includes a number of mammals; notably, we share their taxonomic class, which implies significant genetic relation. If mammals other than humans, such as lemurs, bats, and bears (if included as hibernators), can hibernate, what is stopping us from entering an extended "low-power mode" during cold, harsh conditions?

Some cultures that migrated toward the Earth's polar regions or into high altitudes to access resources may have benefited from entering a hypothermic, low-metabolism, shut-down mode during grim, dark winters. Why suffer at high latitudes or altitudes from poor conditions, shorter growing seasons, dangerously cold temperatures, and long periods with little light?

Fortunately, humanity has triumphed. Our advancing technology, including everything from furs and controlled fire to down jackets and electric socks, continues to help us thrive in cold climates. Despite these triumphs, however, new reasons for discovering a means of human hibernation or hypothermic torpor are appearing on the horizon, motivating some fascinating science. Time and space appear to impose no limits on the current investigations.

## LEGENDARY HIBERNATORS

Although humans are not considered capable of true hibernation, there is a stunning record of people who survived unconscious hypothermic states lasting for days. They stand out as unique after having endured conditions normally considered fatal. These stories originate nearly two millennia ago and continue to the present, each contributing novel insight to current investigations aimed at unlocking the secrets of human hibernation.

Tales of human hibernation, however, begin with the stuff of legend, deep within the realm of fiction. One of the oldest of such tales, if not *the* oldest, concerns the legend of the Seven Sleepers. The earliest of several versions that exist of this story dates back to the turn of the fourth century and was written in Syria. It concerns seven young men who practiced Christianity during the Roman persecutions that were taking place around AD 250. Under threat of being arrested for their monotheistic practice, they were given a chance to reclaim their faith in the Roman gods. Yet they refused. In defiance, they headed to a nearby cave to pray and remained there until they fell asleep in the quiet surroundings. Meanwhile, they had been spotted in the cavern, and word of their whereabouts made it to the emperor, who ordered the cave to be sealed with the Christians still in its cool confines. His commands were carried out without regard for when and how to reopen the cave. The hours turned into days, which turned into months, years, and decades. It seemed their fate had been sealed. Literally.

About three hundred years passed, and Christianity had transformed from a condemned practice into an accepted and even dominant religion, essential to the Roman Empire. The seven young men had been long forgotten. Then one day a farmer decided to open the sealed entrance to the cave. To his astonishment he found the men inside, still sleeping. Incredibly, they awoke to discover a new world in which their religious faith had been consolidated.

Today there exists a grotto in Ephesus, in present-day Turkey, that was excavated in the 1920s, revealing hundreds of graves from the fifth and sixth centuries with inscriptions on the walls dedicated to the "sleepers." The tale, of course, is just that—a fictional story—but its significance lies in an early conception of virtual time travel through hibernation, undertaken in an environment commonly associated with cold.

Throughout the centuries, other, less obviously fantastical (but still questionable) stories have involved mysterious feats of human hibernation—or at least something akin to it. One concerns groups of people who have thrived in cold climates for ages—cultures that have long adapted their lifestyle to the unique conditions found in such environments.

Survival measures employed in notably cold regions can be drastic and even extreme. According to legend, residents of the Pskov region of Russia adhered to a practice of energy conservation called *lotska* that lasted up to six months. Quoting from an article published anonymously in the year 1900 in the esteemed *British Medical Journal*:

Not having provisions enough to carry them through the whole year, they adopt the economical expedient of spending one half of it in sleep. This custom has existed among them from time immemorial. At the first fall of snow the whole family gathers round the stove, lies down, ceases to wrestle with the problems of human existence, and quietly goes to sleep. Once a day every one wakes up to eat a piece of hard bread, of which an amount sufficient to last six months has providently been baked in the previous autumn. When the bread has been washed down with a draught of water, everyone goes to sleep again. The members of the family take it in turn to watch and keep the fire alight. After six months of this reposeful existence the family wakes up, shakes itself, goes out to see if the grass is growing, and by-and-by sets to work at summer tasks. The country remains comparatively lively till the following winter, when again all signs of life disappear and all is silent, except we presume for the snores of the sleepers. This winter sleep is called lotska.

Surprisingly, our anonymous author continues with a flattering tone that indicates a sense of envious admiration:

In addition to the economic advantages of hibernation, the mere thought of a sleep which knits up the ravelled sleeve of care for half a year on end is calculated to fill our harassed souls with envy. We, doomed to dwell here where men sit and hear each other groan, can scarce imagine what it must be for six whole months out of the twelve to be in the state of Nirvana longed for by Eastern sages, free from the stress of life, from the need to labour, from the multitudinous burdens, anxieties, and vexations of existence.

Putting such romantic aspirations aside, if there were indeed a prolonged period of forced or purposeful inactivity, it would likely go hand in hand with a maximization of boredom and eventually, aggravated social tension. Considering that a few days spent together indoors during the holiday season is difficult for some families to handle, six months without anything to do could simply be disastrous if not lethal.

Historically, indigenous peoples in the High Arctic of Canada maintained a minimalist lifestyle during winter, generating a vital role for family activities. To stave off boredom and keep spirits up, singing was a central pastime. Relatives and friends engaged in "song feasts" before heading back to their igloos for a long sleep. There was an extensive list of ditties to learn and to sing: songs for dancing, songs for children, songs for making fun or teasing, songs about the weather, and songs for feasts.

As for *lotska*, the legend has not been verified by evidence, and probably never will be. Six months of nearly constant sleep is questionable for a number of reasons. Our activities and sleep/wake cycles correspond with circadian rhythms in our bodies that are biologically connected to the Earth's rotation—the planetary cycle that underlies day and night. These processes, which have evolved over millions of years, were deeply rooted in our biological makeup well before the migration of humans toward high latitudes. They are controlled by factors that are deeply ingrained and concealed from our awareness. Neural regions that regulate cycles of sleep and wakefulness over the span of a twenty-four-hour day are unlikely to be overridden by cultural practices that cater to life in cold climates. Not to mention that most adult humans, regardless of the number of daylight hours, don't need more than nine hours of sleep, which leaves, on average, fourteen hours of wakefulness, whether we want it or not.

It is possible, however, that during winter, groups living in frigid climates made a concerted effort to gauge activity levels with the nutrition and energy available from existing food supplies, and that such supplies consisted of more than bread, as stores of preserved goods could potentially last for months. *Lotska* partakers likely ate little, slept a lot, kept warm, and passed time in ways that prevented boredom while maintaining positive relationships with family, friends, and community, even if there was nothing superhuman about their ability to sleep, their metabolism, or their tolerance level for low temperatures.

The dubiousness of *lotska* as true human hibernation disqualifies it from contributing to the science of how hibernation works. Yet it serves as an important cultural reference in terms of how thinking about cold and torpor coevolved.

## LADY SLEEPERS

Nearly twenty years after that anonymous entry in the *British Medical Journal*, a report was published in the *German Medical Weekly* that would end up exerting considerable influence on the science of human hibernation via a connection with hypothermia. It was a narrative about twenty-three-year old Minna Braun, who was found in the Grunewald forest on a late October day in 1919, early in the era of the Weimar Republic. On the previous day, Braun, overwhelmed by the turmoil of a recent breakup, had decided to end her suffering. She bought lethal doses of morphine and of Veronal—the first commercially available barbiturate. That evening, it was cold and rainy. Braun wandered into the darkness underneath the forest canopy, planning never to come out. She took the drugs and soon fell into what she hoped would be a wonderful, inescapable narcosis. Without any shelter or way to keep warm, she lost consciousness as temperatures dropped to near freezing.

The next morning she was discovered by some townsfolk who were out for a walk. Although she was unconscious and pale and felt cold to the touch, incredibly she exhibited faint breathing and eye movements—some minor signs of life. Soon she was picked up for transport to the nearest hospital, where a desperate revival attempt had been planned. On the way, however, the worst happened: Braun's meager breaths and barely perceptible heartbeat all but disappeared.

She was rerouted to the morgue.

There, a physician examined her and confirmed her death. According to his examination, Braun's body was showing early signs of rigor mortis, known to occur only after death. Moreover, she failed the "life tests" given to her. These included determining whether the patient exhibited any signs of breathing by holding a feather in front of her open mouth and watching for it to move, or by placing a mirror in front of her mouth and seeing if it fogged. More invasive methods included the wax test, in which the skin was monitored for redness and blistering after hot wax was poured on it. After failing the tests, Braun was placed in a coffin. Her death was attributed to two causes: poisoning and hypothermia.

Fourteen hours later, a police official saw that Minna's cheeks were tinged blue. He also noticed something incredible: movement in her larynx. The physician was immediately called in. After a considerably more careful examination, he noticed faint heart sounds even though he could not detect respiration or a pulse. Although pale, blue, rigid, and without consciousness, Braun was lively for a corpse!

With some extreme luck—and probably some skill on the part of the physicians who treated her—Braun was revived and left the hospital "healed."

The initial failure to find any sign of life in Braun is now attributed to the combination of the sedative effects of morphine and Veronal and the exacerbation of those effects by hypothermia. Essentially, she had sustained a hypothermic sleep narcosis. Her condition mimicked hibernation, although it was only much later that scientists made that connection.

Some decades later, in 1951, a similar event occurred that would have equal influence on the science of human hibernation. This time it involved Chicago's "Frozen Woman," Dorothy Mae Stevens, also twenty-three. (This is beginning to feel like an episode of *The Twilight Zone*. You can almost hear Rod Serling introducing the story.)

It was a blustering day in Chicago, Illinois, with temperatures measuring −12°F (−24°C). At 7:45 a.m., police officers discovered Stevens's body. The local newspaper reported that she was wearing only a spring coat over a sweater and a skirt with stockings.

"I could have sworn she was dead, except all of a sudden, she groaned," remarked one of the policemen.

She was taken to a nearby hospital. When doctors attempted to measure her blood pressure, they found it so weak that any accurate measurement was impossible. Moreover, the thermometers in use at the hospital weren't calibrated low enough to measure her temperature. When a chemist's thermometer was obtained, her body temperature was measured at 64°F (18°C)—the lowest ever recorded in a living human at the time.

After about twelve hours, Stevens regained consciousness. She later proclaimed that she "would never touch another drop of liquor," citing it as the culprit of her cold sleep on the streets.

She and her husband had recently separated, and that night she had attempted to drown her problems in alcohol. Apparently she had passed out on her way home. Unfortunately, the frostbite she acquired during her cold narcosis cost her both legs below the knees along with several fingers. She received much attention and support as local citizens followed her progress through newspaper, radio, and television reports. Her notoriety climaxed when her story was covered in *Life* magazine and she subsequently received an entry in the *Guinness Book of World Records* for having survived the coldest human temperature on record.

The similarities between the Minna Braun and Dorothy Mae Stevens cases stand out, suggesting the possibility of artificially inducing in humans a prolonged state of torpor, even hibernation. Both women—potentially the most extreme examples of survival after hypothermia that had been recorded up to that time—were from urban areas not particularly known for extreme climates and where indoor lifestyles were relatively common.

Both victims had ingested narcotics. Was it possible that the particular drugs involved could somehow have supported an extra-human ability to enter a state of extreme hypothermic torpor without perishing? In Minna Braun's case, according to German doctors Burkhard Madea and Eberhard Lignitz in their 2010 book concerning apparent death, *Von den Maden zum Mörder* [From Maggots to Murder], the combination of morphine and cold, if experienced separately, could have proven lethal, but combined they likely constituted a lifesaving chemical-cold cocktail in which both factors mitigated each other. In Dorothy Mae Stevens's case, although alcohol normally exacerbates hypothermia because it increases the flow of cold blood, in extreme circumstances it has been found to lower metabolic requirements and delay heart failure.

In both cases it seems that cold, in combination with the right narcotic contexts, produced a state at least vaguely comparable to hibernation, yielding life-preserving properties that enabled the women's survival over otherwise lethal durations of cold exposure.

The narcotic-cold connection between the cases of Minna Braun and Dorothy Mae Stevens did not go unnoticed. In fact, when it was realized, it sparked the creation of an entire scientific subfield.

## THE SPACE RACE

Back in 1942, Hubertus Strughold, a scientist working under Hitler for the Third Reich, attended a conference in Berlin regarding recent medical advancements that were relevant to the ongoing war effort. The conference proceedings were kept internal—they would remain hidden from the rest of the world until secret Nazi documents describing the proceedings were discovered and analyzed after the war. The papers revealed that during the conference, a hot topic was the Dachau hypothermia experiments. Unlike the Nuremberg proceedings, however, instead of revealing the extent to which innocent victims were tortured under Hitler's rule, the documents revealed the "progress" that had been made on saving German soldiers, mostly pilots. At the time, Strughold was the director of the Aeromedical Research Institute in Berlin, a component of the *Luftwaffe*.

In the conference minutes, Strughold appears to have commented on the "cold studies," stating, "With regard to the experimental scientific research, but also for the orientation of the Sea Distress service, it is of interest to know what temperatures are to be counted on in the oceans concerned during the various seasons." So although there is no official record of him actually participating in conducting the horrific, torturous experiments at Dachau, in which victims were killed in near-freezing water, he was aware of them.

Despite this controversial past, Strughold is best known as a central figure in the American Space Medicine Association. There, his contributions were so well respected that an award for outstanding achievement within the field was established in his name: the esteemed Hubertus Strughold Award. It was given annually for nearly fifty years beginning in 1963. In fact, he's known as the father of space medicine, and his postwar endeavors in the pioneering field contributed to the achievement of the moon landing. His activities as a functionary of the Third Reich only triggered controversy in 2013, after ongoing investigations into Nazi war crimes revealed the relevant details and connections, leading to the removal of his name from the award.

During his time in the United States, Strughold played a key role in establishing a case for achieving human hibernation for prolonged space travel. He was brought to the country in 1947 as part of Operation Paperclip, a secret program of the Joint Intelligence Objectives Agency. During the program, which ran between 1945 and 1959, special agents recruited more than sixteen hundred scientists, engineers, and technicians from Germany. Some were former leaders within the Nazi Party. The primary purpose: win the Space Race. Indeed, by the time Strughold had been recruited, the Soviet Union was more aggressively, and forcibly, acquiring German experts to achieve the same end. By the time the Soviets were finished recruiting, they had rounded up a grand total of over two thousand specialists.

Strughold ended up in Texas, at Randolph Field, where he was assigned to work at the US Air Force School of Aviation Medicine. There, he considered physical challenges to space exploration that went beyond those associated with "aviation medicine," such as hypoxia (a shortage of oxygen at the tissue level) or the potential of hypothermia after an emergency ejection. He began to explore, in depth, the difficulties that astronauts would face during spaceflight. Shortly after his arrival, he coined the term "space medicine," and the field was born. For Strughold and several other scientists also recruited through Paperclip, the sky provided no limit to the possible achievements of humanity and, primarily, America.

During the Space Race, the realm beyond Earth's atmosphere was seen as the "final frontier"—a new territory to bravely conquer in much the same way as, for example, the Arctic and the Antarctic had been previously, and before that the deserts and the tropics. The latest advancements in science and technology not only rendered human space exploration possible; the moon appeared ripe for the taking. But it wouldn't be easy.

In the context of the Cold War, which was in full swing, the United States saw the moon landing as essential for world domination. Space was a military proving ground; if America couldn't get to the moon, the Soviets—and communism—would, thereby demonstrating technological domination and superior military capability. It was a race *for* the skies, *in* the skies.

With the potential of stratospheric military flight and orbiting missile platforms coming into existence, the threat of Soviet domination was growing increasingly serious. By 1962, the USSR had already claimed three major victories in the Space Race: in 1957, the first circumnavigation of an artificial satellite, Sputnik, around the planet; in 1962, the first orbit of a human, Yuri Gagarin, around the Earth; and in 1968, the successful orbit of the first living creatures—two tortoises—around the moon. Ultimately, in political terms, the inhospitable, cold, distant emptiness of space that had kept the lunar surface beyond the reach of humankind needed conquering, for the sake of freedom and democracy, or for world domination—however you may see it in retrospect.

In any case, conquering cold was going to take some work. In a seminal 1951 paper titled "Where Does Space Begin?," Strughold, along with other Paperclip colleagues, pointed out that fatal temperatures would be encountered at an altitude of sixteen kilometers, considerably lower than the hundred-kilometer threshold that defined outer space. At the sixteen-kilometer boundary, temperatures could get down to −238°F (−150°C).

## "OPTIMAN" OR CYBORG?

Shortly after the National Aeronautics and Space Administration (NASA) was formed in 1958, its researchers came up with various approaches to human space travel, lunar landing, and interplanetary voyages. Within these conceptualizations, engineers and rocket scientists consistently agreed that they would have a vastly simpler job if they could simply eliminate the one factor that repeatedly spoiled all straightforward engineering designs—the human astronaut. Ironically, they were regarded as the weakest link in conquering space.

Humans in orbit for extended durations need a steady supply of oxygen, protection from cold and radiation, a manner of eliminating waste, freedom of movement, orientation in zero gravity, a reliable method of communication with Earth, and a supply of nutrients and water. Comfort aside, on a basic level any successful design would have to cater extensively to keeping its human operators alive and highly functional. And to this day that remains true, whether it be for an hour-long Earth orbit or a potential sixteen-month mission to Mars.

Such considerations caused some engineers to take a reverse approach. To eliminate or abate the human spoil factor, why not model the astronauts for space travel instead of modeling the ship for the astronauts?

Enter Bruno Balke, another German recruit from Paperclip. He was a physiologist with mountaineering expertise—which gave him an edge when it came to surviving in the cold. Balke was inspired by indigenous people from cold climates and high altitudes, such as the Andes and the Himalayas, where residents were thought to have a particularly high tolerance to cold weather. He reasoned that if humans could evolve such abilities, science could be used in place of evolution to achieve the ideal astronaut body, reducing the millennia it would otherwise take to mere years or decades. (As far as we know, he never considered inviting any indigenous people who already had the abilities he sought to enter an astronaut training program.)

Balke's sentiments were echoed in an article coauthored by Toby Freedman and Gerald S. Linder that appeared in a 1963 issue of *Space Digest*. Freedman was director of life sciences at North

American Aviation, and Linder was acting chief of the Aerospace Medical Association. They wrote:

> If we look at the evidence, all over the world, people have developed by hit-and-miss, pseudo-scientific, empirical methods, what are literally superhuman abilities. Tibetan Lamas can maintain normal skin temperature in subzero cold, Yogis buried alive manage on a fraction of normal oxygen consumption, Eskimos thrive on a high-fat diet that would give us all coronaries, Peruvian Indians do heavy labor at altitudes where you or I couldn't breathe. These remarkable adaptations have been accomplished without the benefit of science. Now science, for the first time, has taken the trouble to look into these phenomena and investigate them experimentally. This will almost certainly lead to an understanding and control of acclimatization.

So the question became: How can humans, in essence, be artificially evolved for acclimatization to the cosmos? When it came to the Space Race, humans were considered suboptimal, and if America didn't make them optimal first, the Russians surely would. The solution was to create a superhuman "Optiman" who not only could withstand space's frigidity, among all its other adversities, but could thrive in those conditions by force of sheer physical stature.

The authors continue:

> How would you like to be able to apply a compressive force of several thousand psi, to play six-note chords on the piano, to have fingers with interchangeable screwdriver ends? Nothing to it. The Russians are working on it right now—not a tool, not a machine, but a bio-hand, operated by the owner's own nerves and muscles. We will have the ability to

modify man. . . . We see a man whose outward appearance is quite normal, but who has been adapted to the oxygen requirements of a Himalayan Sherpa, the heat resistance of a walker-on-coals, who needs less food than a hermit, has the strength of Sonny Liston and runs the mile in three minutes flat, while solving problems in tensor analysis in his head. We call him Optiman, and we think we can make him in the near future. If we don't, the Russians will. With our high-thrust rockets we have gone a great way toward conquering space. When we are able to suspend animation for long periods by putting people into hypothermic hibernation, we will have pretty well conquered time. Let us hope that by then we will have modified man in the right direction and have conquered ourselves.

The plan was for the hypothermic, hibernating Optiman to conquer space, perform acrobatics on the moon, and ultimately win the Cold War. What's more, his creation would allow us to evolve artificially, beyond the limits of space and time and of *ourselves*—the limitations of our very human form.

Ambitious indeed.

Focusing on hypothermic hibernation in an article in *Life* that came out the following year, the author, Albert Rosenfeld, quotes Freedman, who conjectures, "What better way to pass the time than in a sort of hibernation? Perhaps we can devise techniques that will enable the Astronaut to slow his heart rate, lower his body temperature, reduce peripheral circulation and then curl up like a wood-chuck."

Okay, now let's step back. Notice that there seems to be a stark lack of comprehensiveness here. It's purely rampant and wild conjecture. Exactly *how* all these adaptations could be achieved seemed to point in one suspiciously nondescript direction—science. And yet the only naysayers to these fantasies were also scientists.

In retrospect, it can be argued that such an extreme version of retrofitting humans to thrive in extraterrestrial environments was a fringe sentiment. But it wasn't the only one. Equally outlandish as a means of conquering outer space, from today's perspective, was the serious consideration of creating—the cyborg!

Now an iconic part of science fiction, the cyborg originated as an entirely realistic aspiration. The term combines *cybernetics*—the scientific study of control and communication between animal and machine—with *organism*. It was introduced in 1960 at a Space Medicine conference by Manfred Clynes, a computational scientist, and Nathan Kline, a psychopharmacologist. Their concept of a cyborg basically comprised an astronaut man-machine that was able to maximize self-sufficiency in space. Indeed, they were proposing another approach to modify the man instead of the ship. The two fleshed out the idea in an article titled "Drugs, Cybernetics and Space: Evolution to Cyborg"—their contribution to a 1961 conference publication titled *Psychophysiological Aspects of Space Flight.*

Clynes and Kline were inspired by a recent cybernetic invention, the Rose-Nelson osmotic pump, an automated drug-delivery device that could be surgically implanted in an organism to achieve an optimal level of drug dosage over a given period of time. In a manner similar to the rationale behind Optiman, they suggested that "man no longer needs to rely on natural evolution and heredity to suit his environment." Rather, via "suitable biochemical, physiological and electronic modification," man would be able to function optimally in the vacuum of space. They wrote, "If a man in space, in addition to flying his ship must be taking continual checks and making continual adjustments in order to keep himself alive, he becomes a slave to the machine. The purpose of the Cyborg is to provide an organizational system in which these robot-like problems are taken care of automatically and unconsciously, thus freeing man to explore, to create, to think, and to feel."

To make their case, they provided an analogy regarding a terrestrial fish, explaining that if a fish desired to walk on land, such a feat could in principle be accomplished if the fish had sufficient *scientific* knowledge to apply to the task, emphasizing that it would not have to rely on any evolutionary processes.

When it came to acclimatizing a human to go beyond the evolutionary boundaries that prevent direct functionality in outer space, scientists focused on what they thought was a problem of *efficiency*. Specifically, the awake human's ridiculously high energy requirements in the forms of food, liquid, and oxygen were a significant limiting factor for prolonged space travel.

Again, similar to the Optiman plan, the cyborg solution involved a radical lowering of metabolism achieved through hypothermic cooling. But there was more: a cyborg's temperature could possibly be controlled by the use of a vascular shunt that would reroute blood into a miniature thermoelectric cooling device and then circulate it back into a vein or artery. Alternatively, drugs acting on the "brain center" that controls metabolism could achieve a more direct control of temperature and wakefulness.

Clynes and Kline introduced the possibility of using a hormone to alter pituitary control of metabolic functioning and to induce a state of hibernation. Even more exciting was the possibility of differential control over brain and body, such that via cybernetic cooling, a cyborg's body could remain in an efficient, cold, zombie-like state while the brain, under the control of various amphetamines, could operate the body in a fully conscious, super-awake state. This alertness, of course, would be controlled so that throughout a flight lasting potentially thousands of years, the cyborg would remain in hibernation for decades or even centuries, while still being able to reactivate consciousness when approaching points of interest, allowing for careful observation and study of astronomical phenomena.

The 1964 Rosenfeld article in *Life* ventured even further into the land of ungrounded speculation:

The Cyborg, though cybernetically controlled, would be a human being—if, after radical tampering, he could still be called that. . . . A Cyborg would still look like a man, but an unearthly one indeed. He would be encased in a skintight suit, needing no pressurization because his lungs would be partially collapsed and the blood in them cooled down, while respiration—and most other bodily processes—would be carried on for him cybernetically by artificial organs and senses, some of them attached to the outside of his body, some of them implanted surgically. His mouth and nose, too, would be sealed over by the suit, because he would not need them to breathe with. Cyborgs would communicate with one another by having the electrical impulses from their vocal cords transmitted by radio. The artificial organs—actually a tiny, complex computer system constantly receiving and feeding back information to regulate the body to its changing environment—would keep a Cyborg's metabolism steady despite radical fluctuations in external temperatures and pressures. The Cyborg could travel in an unsealed cabin through the vacuum of space, walk around on the moon or on Mars protected from heat, cold or radiation by a variety of chemicals and concentrated foods being pumped directly to the stomach or blood-stream. Wastes would be chemically processed to make new food. The tiny bits of totally worthless waste matter would be deposited automatically in a small canister carried on the back.

It seems as though, by 1964, both the Optiman and cyborg approaches to conquering space (and cold) were entirely focused on the human factor, yet neither seemed to have any regard for the *humane* factor. Issues regarding physical comfort, psychological well-being, and a multitude of other risk factors inherent to such drastic modifications of the human form failed to make a blip on any ethical radar.

Going beyond these crazy steps toward transforming "a man" was a potential "giant leap for mankind." Yet while extraplanetary exploits were redefining the limits of human existence, they were also redefining human adequacy. The idea of creating space-hardy superhumans now meant that terrestrial ones were somehow inferior. Meanwhile, the question of whether or not humans *could* be improved—and what exactly defined improvement—were issues open to debate. Both of these weird approaches treaded murky waters surrounding moral and ethical values.

## HIBERNATION AND TORPOR FOR PROLONGED SPACE TRAVEL

Well before these outlandish conceptions of artificially modifying humans for the extreme inhospitableness of space were tossed about, a more down-to-earth approach was initiated by Strughold. He realized that the simple idea of cold as a formidable enemy was actually more nuanced. He believed that harnessing the preservative power of cold could afford advantages that would enable humans to cross extreme distances in space while remaining essentially the same age and in top condition. Directly inspired by the cases of Minna Braun and of Dorothy Mae Stevens, Strughold realized the connection between hypothermia, drugs, sleep, and survival. Indeed, he was the first to suggest using a state of suspended animation, similar to hibernation, in which the astronaut lies cold and essentially dormant while hurtling through the galaxy.

He proposed the idea, indirectly, in a 1954 *Journal of Aviation Medicine* article regarding oxygen deficiency, noting that recent research had shown that cold tissues require less oxygen, explaining that certain animals, such as mice, which notably are mammals like us, are able to go into a "low-power mode" in which they maintain a reduced body temperature and exhibit better tolerance to high altitude. He also described the cases of Minna Braun and Dorothy Mae Stevens and wrote in a final statement, "It is worthy of note that in certain organisms, under certain conditions, the life processes can be suspended completely for an indefinite period, and then resumed. . . . The state into which these organisms fall . . . is known as suspended animation. In all these cases life becomes latent or dormant. Such a state may last for thousands of years."

The implications were profound. The US Air Force took the notion seriously and funded initial research with dogs, exposing them to hypothermic temperatures while they were under sedation. Because of reductions in metabolism and oxygen requirements brought about by the narcotics, some of the dogs survived and recovered. These results offered enough motivation for the research to continue.

By the 1960s it was accepted that some form of artificially induced hypothermic suspended animation could confer a wide range of benefits. These included improved resistance to cold temperatures, high g-forces, and radiation. Additionally, prolonged suspended animation could provide psychological benefits such as reductions in stress, boredom, and claustrophobia and a decreased chance of interpersonal conflicts otherwise likely to arise in a group confined to a small space over an extended period of time.

Surprisingly, and most importantly, hypothermic suspended animation could produce a significant reduction in spacecraft weight. The potential savings were tremendous. According to calculations in a 1963 *Aerospace Medicine* article, for a voyage to Mars, "requirements for food, oxygen and water plus the apparatus necessary to neutralize exhaled carbon dioxide amounts to 2700 lbs./man over the 8 and ½ month period." Although unstated, also implied was the fact that the cost of launching a spacecraft weighing well over thirteen thousand pounds into orbit and to Mars was itself astronomical.

Thus, rather than enemies, cold and hypothermia came to be regarded as allies in the Space Race. What's more, early research showed that for each degree of reduction in core body temperature, the rate of metabolism also decreased significantly. Achieving human hibernation seemed optimal. Yet scientists acknowledged that methods for safely inducing the hypothermic conditions necessary for extended space travel were at the time crude and unsafe; moreover, not enough was known about hibernation to actually apply it to humans.

Then, in 1969, history changed. That year, in front of the largest television audience ever recorded, Neil Armstrong set foot on the moon. It was a monumental accomplishment that marked a significant change in history. Humans had proved themselves capable of conquering the "final frontier" by traveling to the moon and back. The Space Race had been won, and the United States had secured its top world status over the Soviet Union.

Although the Cold War raged on, the Space Race had reached its apex. With the Apollo program having cost up to 4 percent of the US federal budget and the Vietnam War growing in expense, NASA's funding was severely cut. No longer were fantastical ideas for how to achieve interplanetary travel tossed about, let alone funded for research. Although using cold as an integral part of achieving human hibernation for extended periods remained the most realistic approach to traveling distances on the scale of a Mars trajectory, for decades any direct action was put on hold. The lunar landing was viewed as a "mission accomplished."

◇◇◇◇◇◇

THE TWENTY-FIRST CENTURY, however, has seen a remarkable and ambitious revival of interest in both government-funded and privatized space exploration. The research firm Bryce Space and

Technology reported that so far within the new millennium, $18.4 billion has been invested in commercial space ventures. Private companies like SpaceX aim to create a colony on Mars as early as 2025. Presently, the company is beginning to turn a profit by launching satellites, delivering goods to the International Space Station (ISS), and offering commercial spaceflights. Anyone up for a $55 million trip to the ISS?

There's also renewed interest in basic scientific discovery. The moon, Mars, and asteroids offer insights into the origin of the Earth, our solar system, the galaxy, and the universe. Moreover, they offer insight into the origin of *life* on Earth and into whether alien life exists. What humans can accomplish on the moon and on Mars by actually being there is still considerably greater than what can be achieved by sending sophisticated probes like the Mars rovers. Government-funded projects aimed at sending astronauts to the moon, to Mars, and even to asteroids are in the works. And with agencies like NASA operating at around only 10 percent of the budget they had in the 1960s, technological advances in artificial intelligence (AI), machine learning, sensor technology, and robotics are putting space exploration back within financial reach. In the 2020s, NASA plans to return humans to the moon (2024), to create a new ISS habitat for astronauts, and to eventually send humans to Mars. India plans to put its first astronauts in space (2022), and China plans to put humans on the moon in the 2030s.

Just as they were during the Space Race, considerations regarding weight, the adverse conditions of lengthy spaceflight, and the benefits of a hypothermic state of suspended animation are cornerstones of ongoing research into the goal of a manned flight to Mars. Key advances have surfaced from cases of prolonged accidental hypothermia that are even more extreme than those of the twentieth century.

One in particular stands out—that of Mitsutaka Uchikoshi, a truly remarkable story of a man who, after sustaining an accident outdoors, ended up in what is widely considered a state of suspended animation that doctors agree was comparable to hibernation. Phenomenally, this case reverses dogmatic thinking on the impossibility of human hibernation without any sort of pharmaceutical intervention, unlike the past experimentation involving "frozen sleep" or sleep narcosis.

On October 6, 2006, Mitsutaka Uchikoshi attended a barbeque with some colleagues on Mt. Rokko in western Japan. Employed as a civil servant in Kobe, Uchikoshi, age thirty-five, was in generally good physical condition with no significant health problems. After the gathering, rather than joining his associates on the cable car for the trip home, Uchikoshi decided to navigate his own return by walking down the three-thousand-foot mountain alone. Suddenly, while descending, he slipped in a stream and fell. Struggling, he made his way out of the stream but realized he was injured; he had broken his pelvis. Unable to walk, Uchikoshi was stranded and had to wait for help. After some initial optimism, he realized that the minutes were turning into hours. And the hours into days. The last thing he recalls of his experience on the mountain is falling asleep on the second day. That night the temperature dropped to 50°F (10°C).

Twenty-four days later, Uchikoshi was discovered by a passing hiker; astonishingly, he was still alive but unconscious. He was taken to Kobe hospital, where his core body temperature measured a mere 72°F (22°C). He showed signs of organ failure and blood loss. Incredibly, he fully recovered.

Doctors surmise that when Uchikoshi fell asleep, his body entered into a hypothermic state, lowering his metabolism. The reduced metabolic activity apparently rendered him capable of surviving for weeks without any food or liquid. Furthermore, he seems to have remained neurologically unaffected. During hibernation, as metabolism is reduced, breathing grows less frequent, and less oxygen is carried in the blood to the brain, which otherwise highly demands it. Intact neurological functioning also depends on normal temperature conditions for a host of other reasons involving biochemical and electrical properties.

Essentially, Uchikoshi hibernated. He's the only recorded case of someone to survive for so long in a state of unconscious, deep hypothermic torpor without a source of nutrients. His story, widely covered in the press, created a stir in the medical community. Many doctors voiced skepticism of the reported events and whether or not he was left without lasting neurological damage.

Upon his recovery, Uchikoshi simply commented, "Sorry for all the trouble I've caused."

Assuming that everything reported about Uchikoshi *is* accurate, his case marks the difference between zero and one—existence and nonexistence—for an elementary state of human hibernation. It implies that extended human torpor in a hypothermic state can be achieved unaided by artificial means for at least twenty-four days, possibly longer. This makes a promising case for exploring human stasis in the context of prolonged space exploration.

Yet we can't go breaking astronauts' pelvises before takeoff and turning down the temperature to potentially fatal levels and simply hope for the best when they approach Mars! Although Uchikoshi survived the greater part of a month in a state of unconscious torpor, with a body temperature hovering around 72°F (22°C), he sustained multiple organ failures, and his neurological outcome was questionable. Thus, he's not a perfect model, as highly motivating as his case is.

But Uchikoshi's experience isn't the only event of the last few decades to have significant relevance. A similarly extreme case occurred in a controlled clinical setting but with mild rather than deep hypothermia. In 1996, a patient in China with severe traumatic brain injury was treated with cold for a staggering fourteen days. And in 2009 another patient with a brain injury—this time nontraumatic—was treated in the United States with mild and moderate hypothermia, also for a staggering two-week period. While chilled, these patients were under sedation and were carefully administered paralytic drugs to prevent shivering. Both not only survived their treatments but experienced positive outcomes.

These two cases do mirror the narcotically mitigated survival of Minna Braun and Dorothy Mae Stevens, although neither of them was an accidental hypothermia case, the hypothermia was less severe, they were chilled for considerably longer, and the overall outcomes were better. The key conclusion is that under careful control, and with the right pharmaceutical combination, it is possible to keep someone in a living, breathing state of cool, unconscious narcosis for a prolonged period, in line with definitions of suspended animation.

The prospect of using an extended state of torpor in astronauts for long space journeys is now, again, being seriously considered to make a trip to Mars feasible within the next decades. The benefits conferred by a hypothermic suspended animation, during which astronauts would remain still and unconscious, are worth pursuing mainly because they would result in significant weight reduction, which translates into reductions in propulsion requirements and ultimately cost. In terms of the amounts of nutrients, water, and oxygen required by a single human, and the amount of waste generated, the figures remain in the same ballpark as they were during the Cold War–era Space Race. True, improvements have been made in our ability to extract water from waste and to more efficiently and effectively provide nutrients, but humans' caloric needs haven't significantly changed. For a mission to Mars, they translate into about two thousand pounds of payload per person. Propelling that mass out of the Earth's orbit and to Mars for a minimal crew of three remains a significant engineering obstacle. Moreover, at $10,000 per pound, the cost literally becomes prohibitive—$66 million at the very least. That's not even considering the return trip (although caching supplies at the destination has been considered).

## HIBERNATION OR TORPOR?

Broadly, what characterizes hibernation as a form of suspended animation—relative to torpor, for example—is that it is self-initiated. Some process, still unknown, triggers an alteration in physiological function that reduces heart rate, respiration rate, immune function, and digestive function—and consequently waste production. The lowering of body temperature is a natural result, primarily from the reduction in metabolism. How to induce hibernation in humans? That is *the* question.

Hibernation is one of the main components of a Mars voyage, as envisioned by the European Space Agency. In its current conception, the plan stipulates that the astronauts would first beef up, eating extra calories to acquire body fat, just as hibernating nonhuman animals do before settling down for months. Then, upon initiation of the voyage, while being carefully monitored in small, individual pods, they'd receive hibernation-inducing pharmaceuticals. The pods would darken and reduce in temperature to facilitate cooling. Since communication with Earth could entail up to a twenty-minute delay, the astronauts would be monitored primarily by advanced technology including AI in addition to fault-detection, isolation, and recovery systems. The hibernating crew would awaken about twenty days in advance of their arrival for an adjustment period, during which their pods would double as cabins. Those three weeks would comprise the final segment of a 180-day voyage to the red planet.

So, how to achieve this? Inducing human hibernation will likely require a deeper understanding of hibernation in other animals—particularly in mammals: groundhogs, lemurs, squirrels, hedgehogs, and, for those brave scientists out there, bears. Studying their genetic code could hypothetically give us a clue to the genes that trigger the specific physiological sequences that initiate, maintain, and terminate hibernation. What do we know so far about the specialized genes that enable them to do this?

Nothing!

Why? Because as of yet, there doesn't seem to be anything particularly unique about their genetic code in terms of enabling hibernation. Assuming this remains the case, it is a crucially important discovery. Surprisingly, sound genetic evidence implies that the ability to hibernate could have developed not with the evolution of new species, but rather with the very first ancestral mammal. This inaugural mammal was a tree-dwelling, shrew-like creature that roamed the planet around 160 million years ago. Since it gave rise to all modern mammals and therefore to all mammalian genetic hardware, including ours, hibernation may indeed be an executable form of human genetic software. Otherwise, it appears attainable in nonhibernators via some relatively minor genetic tweaking. Thus, a remote possibility exists that whatever triggers a cascade of hibernation-producing physiological reactions in them could do the same in us under the right conditions.

What are these conditions? For most hibernators, their stasis is spurred by environmental factors, like the onset of winter, that cause a loss of food resources and are best waited out. In fact, hibernation is so tightly linked with food supply that many hibernators kept in zoos don't hibernate, because resources there are consistently plentiful. But exactly what indicates a looming lack of resources and triggers hibernation? If not weight loss, is it the onset of colder temperatures? Perhaps a reduction in daylight duration? A decline in food or water within their environment, signaled somehow by increasingly infrequent meals? Or maybe some combination of these? As of now, we don't really know.

In other hibernating animals, the metabolic reduction that occurs in cold, harsh conditions is an integral part of life. For them, the physiological processes that initiate hibernation are entirely controlled by internal rather than external triggers. These animals take no chances; they'll go into stasis no matter what. Researchers studied golden-mantled ground squirrels by keeping them inside with a ready supply of food, at a constant temperature, in cycles of twelve hours of light and dark, for three years. The creatures had no exposure to any sort of environmental cue to suggest that winter or drought—a harsh condition—was coming. They hibernated each year anyway. Their hibernation was therefore likely initiated by what are called *circannual* rhythms: biorhythms connected to the Earth's rotation on its axis and that occur over the span of an entire year, in line with the Earth's orbit around the sun.

One thing we do know is that hibernation is often initiated during deep sleep. In humans, sleep is associated with a reduced metabolism and a lower body temperature. Hibernators, however, slow their metabolism considerably more and thus reduce their temperature to levels well below what would initiate a shivering reaction in humans. And since humans automatically regulate their temperature by shivering when they get too cold, any hibernation trigger would have to switch off this response because shivering increases metabolism and body temperature, which would defeat the purpose of hibernation by creating energy inefficiency.

Another obstacle, which may at first seem ironic, is sleep itself. Although hibernation is initiated during sleep and thought of as a period of very prolonged snoozing, it's actually not. The defining feature of hibernation is energy conservation, in the form of slowed-down life-supporting functions. For example, dwarf lemurs—the only known hibernating primate—slow their heart rate from three hundred beats per minute during waking life to only around six beats during hibernation. In correspondence, their respiration is reduced from a breath every second to one every ten minutes. Brain activity, which characterizes sleep and is essential for restful sleep, is reduced to nearly zero.

Surprisingly, this means that hibernators are likely to become very sleep deprived. To remedy this problem they self-initiate an arousal process that modifies the brain from an inactive hibernation mode to a more active sleep mode. Evidence suggests that they *arouse* to sleep because they have accumulated as large a "sleep debt" as they can sustain. During these arousals, they re-warm themselves as their heart rates and respiration rates return to around normal levels and their brains exhibit increased patterns of activity characteristic of sleep. They do this relatively quickly in comparison with the rewarming processes carefully administered to hypothermic patients. In humans, the process is undertaken slowly for fear of increasing the circulation of cold blood, which would result in further, possibly fatal cooling.

Finding a way to engage a similar arousal process that brings an astronaut out of the extreme, inactive state of hibernation into a more active sleep state in order to meet sleep requirements poses a significant problem for using hibernation for prolonged space travel. Moreover, the increased energy requirements of the more active state of sleep could translate into larger payload require-ments, which goes against the whole point of using hibernation in the first place. If this isn't resolved, is there any other option?

There is. And it is more in line with the cases of Minna Braun and Dorothy Mae Stevens, which got the ball rolling in the first place. The approach involves inducing and maintaining a state of hypothermic torpor rather than full-on hibernation.

Ongoing research funded by NASA is underway at Space-Works, an aerospace outfit based in the state of Georgia. Their approach is to keep a rotation of astronauts in a low-metabolism, hypothermic condition by maintaining a chilled ambient tempera-ture within a pod-like containment unit. The astronauts would re-main under sedation while intravenously receiving nutrients in a manner that does not require regular, waste-producing digestive processes. To prevent muscle atrophy, and to keep bones strong despite lengthy stretches of inactivity, they would be electrically stimulated by a robotic system. At any given time, one astronaut would be awake, monitoring the remaining crew, communicating with Earth, and maintaining the ship.

Compared with a previous NASA design for a Mars voyage, SpaceWorks estimates reductions in astronaut habitat mass ranging from 52 percent to 68 percent, figures that significantly increase the possible number of crew members.

In the company's current conception of an interplanetary spacecraft, the primary habitat consists of "sleep chambers" in which the sedated astronauts are held in place with zero-gravity restraints while chilled to a core body temperature of between 90°F and 82°F (32°C and 28°C). Temperature adjustment is enabled by bedding pads that include warming circuitry. Vital signs, including measurements of dehydration based on urine output, are monitored by a computer. Using a system similar to those currently in place for long-term-care patients, intravenous nutrient administration occurs automatically and bypasses regular, energy-intensive digestive processes. The astronauts would receive a nutrient cocktail that includes dextrose, amino acids, lipids, vitamins, and trace elements. Waste is recycled using a filtration system.

Ideally, the stasis period would last at least two weeks. Because this is the maximum duration patients have successfully been kept under sedation in a hypothermic state, it represents the minimum cycle period for torpor on a mission to Mars.

So, which might be a better solution: synthetic hibernation or torpor under sedation? The body during hibernation is self-maintained, and therefore measures to regain consciousness would initiate from internal processes, ideal temperatures could be maintained automatically, and shivering would not be a major issue. Yet the approach for inducing torpor is considerably more developed. It is based on existing medical knowledge, practice, and technology. For example, current electrocardiogram systems for monitoring cardiac activity and other vital signs for patients in comas or in extended care would need little modification. Same goes for a cooling system. Moreover, it does not require any major genetic discoveries or advances in existing knowledge. It's likely that testing in an actual in-orbit environment may happen within the next decade. Hibernation, however, involves more unknowns.

Nonetheless, some major risks accompany the torpor approach. Humans haven't evolved to sustain extended periods without consciousness or muscular activity. And, as a stark reminder, hypothermia is a highly adverse state of being! As of yet, the officially recognized risks of an enduring state of torpor include blood clotting and infection due to prolonged IV use, decreases in blood coagulation, electrolyte imbalances due to hypothermia, and liver complications due to unusual nutrient sources. Also, as a result of prolonged periods in a microgravity environment, astronauts risk losses of muscle and bone density, diminished eyesight, increased intracranial pressure, spinal elongation, and altered immune-system functioning. Not to mention the dangers associated with increased radiation outside Earth's protective atmosphere.

When we do finally figure out how to keep Mars-bound astronauts chilled, they will be sent hurtling through space at speeds of around twenty thousand miles per hour. They probably won't be conscious until they wake up, so they won't remember their months-long interplanetary voyage. Not only will hypothermia have slowed their molecular structure; it will have slowed their sense of the passing of time.

But there's also a literal time-slowing aspect to their journey, and it has to do with physics. According to Einstein's theory of special relativity, objects in motion relative to other objects can undergo time dilation—the slowing of time. In fact, the effect was tested with accuracy in 1976 by using atomic clocks, one on Earth and the other placed into orbit at an altitude of ten thousand kilometers and accelerated at high speed. When the two clocks were compared, sure enough, the one sent into space had ticked off less time. The implication is that, as Mars explorers are propelled toward the planet at high speed, time will pass more slowly for them than for those experiencing time on planet Earth. The time difference will be doubled on the return home.

Time dilation was observed between identical-twin astronaut brothers Scott and Mark Kelly on a NASA mission aimed at studying the effects of prolonged space travel. Mark remained on Earth while Scott spent eleven continuous months in the International Space Station orbiting the planet a total of 5,440 times. Upon his return, Scott remarked that because of the effect, he "came back from space younger" than his twin.

A Mars-bound vessel would likely travel faster than the ISS and would thus increase the effect of the relative time difference. How much would the total savings amount to?

No more than half a second! Although it is a fascinating reality of space travel, the scale of relevant time dilation is trivial. Nonetheless, the idea of an inherent connection between time travel and suspended animation supported by cold remains consistent in famous fictional stories. Literary and cinematic works like *Planet of the Apes*, a number of the *Alien* films, *Avatar*, *Austin Powers*, *Captain America*, and *Buck Rogers* include heroes who emerge from decades or centuries of suspended animation and yet remain completely unaged.

◇◇◇◇◇◇

*You're so cold now that you couldn't have previously imagined how this feels.*

*Again, you see objects, forms in the distance. But this time it's different—you're sure about it. You study them, and the more you look the more you are confident that what you see are people walking. You get up and try to run, but after only a few steps you stumble and fall.*

*You look up but no longer see the figures. Still, you conjure the energy to shout, "Heeeeeyyyyy!"*

*You hear a response! You can't make out any words, but you answer. Your response is nonsensical: "Let's try surviving out here for a while so we can all eat later. I'm going to need my feet!"*

*You wait for another response. But suddenly you're distracted by the incredible realization that you no longer feel cold. Rather, you feel a mysterious but wonderfully warming heat. And it's coming on fast. It is splendid. Where is it coming from?*

*Why question it? This could be your salvation.*

*It's getting hot!*

*Without thinking, you take immediate action. Time to dress down. You remove your jacket.*

*No difference.*

*To hell with it—why not just take everything off? It would be glorious—like going to the beach and running into cool water on a hot day.*

*You do it. Right down to your birthday suit. But to your chagrin, you still feel uncomfortably hot.*

*At least now you can rest. You lie down on top of your clothes.*

*You feel confident the figures you saw earlier will return to rescue you. The assurance is bliss. You shut your eyes.*

# 5

# BETWEEN LIFE AND DEATH
# AT THE 59°F (15°C) BARRIER

## COLD REVEALED AS AN
## EXISTENTIAL SHIT-DISTURBER

Throughout most of our evolution, from when life began, death and the possibility of an afterlife weren't conceptions on the table for contemplation. We simply didn't have the capacity. Death was simply the end of one's existence. Anthropological evidence indicates that humans began to have religious beliefs and to conceive of life and death relatively recently—between forty-five thousand and two hundred thousand years ago. These conceptualizations were essential to major shifts occurring in cognition and behavior that marked an era in human development known as the "great leap forward." These shifts, rooted in the capability for complex, abstract thought and language, have come to define humanity and behavioral modernity.

Arguably, another existential conceptualization has come about even more recently, within hundreds, not thousands, of years. Involving an ill-defined intermediary state between life and death, it is often referred to as "suspended animation." Although many historical sources provide context for its inception, a significant one involves some rather specific events that were recorded in the seventeenth century. Only now, however, do we realize that cold was a central character in these events.

## FROZEN TO LIFE

Rather modestly, they surround the life of a young woman named Anne Greene, a twenty-two-year-old maid in Oxfordshire who worked for a man named Sir Thomas Read. According to historical accounts, she had an intimate relationship with Jeffery Read, the teenage grandson of Sir Thomas. Apparently Anne said that she "was often solicited, by fair promises and other amorous enticements." She "at last consented to satisfy his unlawful pleasure," an act in which she and Jeffery conceived a child.

Greene had been pregnant for around six months when one day she began to feel pain while mixing a vat of malt. She went to the "House of Office" (outhouse), where she gave birth prematurely to a stillborn child. Horrified, she buried the small body. Soon after, however, it was discovered. Anne was accused of infanticide.

She was sent to prison, and her punishment was decided by the court: public execution by hanging. The date of her hanging was set for the cold morning of December 14, 1650.

Hangings were more gruesome than you might think. Unlike scenes from present-day films depicting a quick death, real hangings were often considerably more drawn out. It was customary for friends and relatives of the condemned to try to hasten the death by pulling on their feet during the hanging. In a comprehensive 1651 account of Greene's hanging, titled *Newes from the Dead*, clergyman and Oxford scholar Richard Watkins writes that Greene "was turned off the ladder, hanging by the neck for the space of almost half an hour; some of her friends in the meantime, thumping her on the breast; others hanging with all their weight upon her legs, sometimes lifting her up and then pulling her down again with a sudden jerk." According to this account, their actions were so disruptive that they were reprimanded and ordered to stop.

When Greene was deemed dead, she was taken down from the gallows and placed in a coffin in a cold area. After about half an hour, her hypothermic body was presented to William Petty, an Oxford anatomist and surgeon, who intended to prepare Greene's body for dissection, which was, at that time, legally practiced on criminal corpses. Incredibly, however, he detected a faint breath. It was followed by another, and another.

Once they confirmed the infrequent respirations, Petty and his colleague, Thomas Willis, began a lengthy and intensive attempt to revive and rewarm her. Their multipronged strategy to revive her involved several of the latest medical "advances": tickling her throat with a feather; administering a tobacco-smoke enema to stimulate respiration; and an old-fashioned blood letting, which was more or less thought of as a humor-rebalancing cure-all. As for rewarming, she was rubbed, given hot medicinal spirits, laid in a prewarmed bed, given a heating plaster, and finally "cuddled up" with a warm, rotund female volunteer.

After all this, one wonders what may have caused Greene more harm: the hanging or the revival treatment. Astonishingly, she went on to a full recovery and was shortly able to return home. At the time, the law stated that if you survived a hanging, it was the result of divine intervention. Greene's innocence was miraculously asserted by the very hand of God. With a smack of morbid humor, she was offered her coffin, which she accepted as a rather tongue-in-cheek memento of the event. Her story drew great interest, and she became a spectacle—someone who many believed had come back from the dead. She was greeted in public by crowds of curious onlookers who were awe-stricken upon witnessing a living miracle. Cleverly, Petty and Willis suggested that Greene ask the spectators to make donations so she could pay for her lodging and for the efforts the two doctors had made to save her.

Despite Anne Greene's remarkable recovery, sadly she died five years later while giving birth—in a rather twisted, unfortunate irony. However, she has lived on, mostly in historical accounts, in medical texts, and as inspiration for literary fiction. Now, rather than assuming she was resurrected by the hand of God, scientists theorize that frigid conditions, perhaps in combination with the blows to the chest she received, may have played an essential role in keeping her alive after her hanging. A 2009 article from the *Journal of Medical Biography* states, "A combination of low-body temperature and external (pedal) cardiac massage after her failed execution, it is suggested, helped to keep her alive."

In a hypothermic state she required relatively less oxygen. Consequently her respiration and heartbeat were reduced and went undetected, especially because she was already taken for dead.

But was her return an instance of resurrection, reanimation, revival, resuscitation, or just plain survival? Although medical knowledge wasn't advanced enough to reveal that she was never actually dead, Watkins's motivation for writing his account was to some extent a desire to remove the supernatural elements attributed to her "resurrection." Rather than declaiming her reanimation a miracle, Watkins's account focused more on the extent to which medical care was given by Oxford medics—care that resulted in a successful revival.

Greene's story came at a time when medicine was becoming less of an art and more of a science. No longer were students learning primarily from the texts of Galen and Hippocrates. Instead, their education was increasingly sourced from books published within their own lifetime. The practice of *experimental philosophy*—scientific experimentation—was becoming central to discovery, understanding, and knowledge. The experimental philosophers at Oxford, where Greene's body was sent to be dissected for exactly those purposes, were beginning to conclude from her mysterious survival that life and death as typically defined were perhaps not the only modes of existence.

Although they had no term for it at the time, these Oxford medics were conceptualizing a state of suspended animation. In this mode, an otherwise living organism can cease fundamental life-supporting processes, for durations normally considered to impose permanent or fatal consequences, including, for some species, heartbeat and respiration. Yet rather than resulting in death, the organism reanimates as its vital processes mysteriously return. Organisms undergoing suspended animation have been alternately identified both as living and as dead but returning to life. Cold is often crucial. Ultimately, through its fundamental deadening of atomic motion, cold is directly linked with slowing the life-supporting processes of heartbeat, respiration, metabolism, and brain activity to a point indistinguishable from clinical death.

Suspended animation, in essence, was first observed by the English physician and experimenter Henry Power as early as 1663. In a series of exploratory experiments on nematodes, or "little white worms or snigs" (as he referred to them), that were commonly observed in putrefying vinegar, Power remarked on their ability to regain life after freezing: "All my little animals made their reappearance, and danced and frisked about as lively as ever. Nay I have exposed a jar-glass full of this vinegar all night to a keen frost, and in the morning have thawed the ice again, and these little vermin have appeared again and endured again that strong and long conglaciation without any manifest injury done to them; which is both a pretty and a strange experiment." Although Power found the creatures "most remarkable," he seems to have missed any profound implications regarding vitality recovered after freezing.

Power's findings, however, were essential to cold science because they initiated a centuries-long succession of attempts by groundbreaking researchers to reanimate various organisms after they had been frozen. These experiments, undertaken to reveal fundamental aspects of life and death, were inspired by a plenitude of accounts regarding the revival of wild organisms, mostly fish, from seafaring laborers, fishermen, and even Arctic explorers. Even today, such experiments continue. As recently as 2018, scientists successfully revived forty-one-thousand-year-old roundworms, similar in kind to the worms studied by Henry Power, from the Siberian permafrost.

## MAD SCIENCE

Around the same time, while Henry Power was making pretty and strange observations of little worms in rotting vinegar, a man named Robert Boyle was coming to prominence as a natural philosopher (the term then for a prototypical scientist). Now regarded as the father of modern chemistry, Boyle was a young academic at Oxford—an anatomy student under the instruction of and well acquainted with the professors appointed to dissect Anne Greene's body. These were inaugural members of the newly formed Royal Society of London. Knowing of the seemingly miraculous conditions surrounding Greene's revival, Boyle embarked on a series of experiments in which he made crude attempts to freeze and revive fish and frogs. He cooled them in water until most of the liquid was frozen around them. Whether the creatures were actually frozen, however, came down to subjective opinion rather than any form of measurement. In any case, they didn't survive. Still, despite the failure of his attempts, Boyle's conclusion on the matter has been cited as expressing that, in principle, even mammals could be revived after freezing.

Wild speculation aside, what *is* significant and remarkable about Boyle's understanding of the connection between cold and preservation was accurate on a fundamental level. He correctly attributed the effects to the slowing of atomic motion. Insightfully, he surmised, "Whilst bodies remain frozen, the cold probably, by preventing that irregular motion and avolition of their particles, whence corruption proceeds, may keep its pernicious effects from appearing."

His observations regarding the movement of particles were influential. They supported a growing branch of philosophy called *mechanism*. Unsurprisingly, it had everything to do with mechanics and held that all natural phenomena could be explained by physical principles governing matter and motion. According to *universal mechanism*, as it is commonly referred to, the universe is entirely physical and material. The proposal left no room for the occult or spiritualism. As surely as the physical principles that govern the motion of a clock's gears can be used to predict that it will chime every hour, the same deterministic principles could be used to predict the physical configuration of anything at *any* time. According to mechanism, if one had the computing power, one could precisely determine every future event in exact, atomic detail.

Crucially, nonhuman animals and human beings were not immune from these physical laws. So not only did Newton's famous laws of motion apply to planets, because he was a physical being himself, they applied to him—or to anyone for that matter. According to universal mechanism, a living being was a sort of complex machine governed entirely by physical and chemical properties, nothing supernatural. Mechanists held that life and motion were intrinsically connected, and so, motion was a sign of life. Cold, as a physical process, was therefore fundamental. If cold slowed atomic motion in a working organic system, it made perfect sense that once motion was restored with heat, the "life" could continue. Suspended animation explained. Done. Simple.

Although mechanism was gaining popularity, some explorers remained skeptical of the idea that life could be explained entirely by physical phenomena in such a refined manner. Henry Baker, a British naturalist and microscopist, explicitly and clearly regarded the reanimation of nematodes and other microscopic creatures as a mystery.

In his 1753 treatise *Employment for the Microscope*, he wrote, "We find an Instance here that Life may be suspended and seemingly destroyed . . . and yet after a long while Life may begin a new to actuate the same Body. . . . What Life really is seems as much too subtle for our Understanding to conceive or define as for our Senses to discern and examine."

These sentiments were expanded by Lazzaro Spallanzani, an Italian priest, biologist, and physiologist, who also explored life as it existed under a microscope. He began cold-centered experiments in 1776, successfully reviving eelworms after freezing them to as low as 0°F (−18°C). The reanimation of organisms, even though they were minuscule, he believed, showed that resurrection did not necessarily require divine intervention, but perhaps wasn't an entirely physical process either. He wrote, "An animal which revives after death is a phenomenon as incredible as it seems improbable and paradoxical. It confounds the most accepted ideas of animality; it creates new ideas."

Indeed, an impulse was forming to explore and reveal the essence of life as it might exist beyond explanations offered by materialist doctrines. The discovery of *anabiosis*—a state of suspended animation—in minute creatures at the threshold of naked-eye visibility, spurred a series of investigations to determine what other creatures could be frozen and revived. Researchers attempted to determine the nature of suspended animation and thereby the nature of life—not only in worms, frogs, and fish, but in humans.

## JOHN HUNTER'S SECRET PLAN

An exceptional, pioneering explorer of suspended animation who held these aims in mind was the Scotsman John Hunter. Widely known today, he was a famous and controversial physician and anatomist when he lived, in the eighteenth century. An advocate of medical experimentation and surgery, he was one of the most distinguished intellectuals of his time. And although some of his investigations in retrospect may seem like hack jobs compared with today's science, during his time, Hunter's experiments with cold reached new levels of sophistication. His work with cold deserves some contextualization.

Born in 1728, by age twenty Hunter was on his way to becoming an expert in anatomy. He served as an army surgeon, in which role he assisted in tooth-transplant operations, a cutting-edge, controversial procedure at the time. By 1764 he had set up his own anatomy school, where he began a private practice in surgery. His interest in anatomy turned into an unparalleled obsession. At his home, he kept an extensive collection of exotic animals—basically, his own zoo. He housed zebras, buffaloes, and mountain goats, to name a few. Over time he also filled the estate with more than fourteen thousand anatomical preparations. They included obscure oddities such as a range of giant tumors, the skull of a boy with a nearly complete second skull conjoined to the back of the scalp, and, his pièce de résistance, the skeleton of Charles Byrne, a "giant" who stood just over seven feet seven.

Inspired by his worldly collection of mysterious, exotic forms, Hunter set out to discover the root of all these various configurations—the essence of life. Initially, and seemingly modestly, he continued tooth-transplant surgery but in new, experimental ways that were actually quite freakish and grotesque. For example, by 1778, he had successfully transplanted human teeth into roosters' combs. In his *Treatise on the Natural History and Diseases of the Human Teeth*, published that year, he wrote:

> I took a sound tooth from a person's head; then made a pretty deep wound with a lancet into the thick part of a cock's comb, and pressed the fang of the tooth into this wound, and fastened it with threads passed through other parts of the comb. The cock was killed some months after, and I injected the head with a very minute injection; the comb was taken off and put

into a weak acid, and the tooth being softened by this means, I slit the comb and tooth into two halves, in the long direction of the tooth. I found the vessels of the tooth well injected, and also observed that the external surface of the tooth adhered everywhere to the comb by vessels, similar to the union of a tooth with the gum and sockets.

Hunter believed that all living material, whether it be a human body, a tooth, a rooster, or its comb, contained a fundamental and mysterious property that gave it life—the "living principle." According to Hunter, "all flesh is the same flesh."

His beliefs were in line with a philosophical perspective known as *vitalism*. It held that there was an essential life-giving property common to all animate beings and proposed that this vital essence existed not only in every creature but also in every part of their physical composition. Hunter held the conviction that "every individual particle of the animal matter, then, is possessed of life, and the least imaginable part which we can separate is as much alive as the whole." This "living principle," as it defined vitalism, provided a complexity beyond the principles of mechanistic philosophy, which proposed that there was no difference in kind between animate and inanimate matter, organic and inorganic. In his *Treatise on the Blood*, Hunter stated, "The mere [physical] composition of matter does not give life, for the dead body has all the composition it ever had."

So if the mere physical/material composition of matter wasn't enough to define a living being, what did? According to Hunter's writing, the answer occurred to him in 1757, while he was studying chicks in the process of incubation. He realized that after an egg was laid, the chick inside, even at the embryonic stage, while seeming completely inactive, was indeed *alive*. Hunter felt that mechanistic notions of movement as a defining property of life could not explain this motionless, embryonic stage of chicken existence. This life continued until the chick began to move and break the eggshell.

Yet while some eggs hatched, others remained still. He noted, with a nod of satisfaction, that "if the egg did not hatch, I observed that it became putrid nearly in the same time that other dead animal matter does."

Hunter realized that perhaps life had something to do with self-preservation, which seemed to be an essential property. Basically, he thought, as long as any living organism wasn't rotting, it was alive. He believed that this property of life via self-preservation enabled an organism to retain what he called "simple life," independently of action. In his *Lectures on the Principles of Surgery*, he wrote, "I have asserted that life simply is the principle of preservation in the animal, preserving it from putrefaction."

He thought that "animal heat"—which animals naturally seemed to retain—wasn't necessarily a life-preserving property, as many others thought it was. Rather, he believed it could be aversive in that it aided decomposition. This was his explanation as to why Blagden and the other men didn't cook in the heated room: "the body has a power of destroying heat." He wrote, "I own I had formed an opinion on this subject. I rather supposed that animal heat was owing to some decomposition going on in the body, and in pretty regular progression, though not the process of fermentation."

Hunter believed that these essential properties could not be explained by physical and mechanical principles alone. Rather, underlying such mysterious functioning was the action of an enigmatic, incorporeal "vital principle." This reasoning led him to believe in the possibility that life could be maintained during freezing, and in the possibility of reanimation upon careful thawing and rewarming. He thus entertained the prospect of suspending animation indefinitely. With eagerness, he went to great efforts to freeze and thaw a variety of animals and organic material to see if they could retain vital signs—and therefore life. It became somewhat of an obsession that would occupy him for years.

In 1766, Hunter set out to prove that when an animal was induced into a state in which it could not produce any action or even its own animal heat—meaning when it had lost all metabolic function—it could still survive. To begin, he procured a glass vessel, a "freezing mixture" containing sea salt and spirits, and plenty of snow. To this he added two healthy, unsuspecting carp. With everything in the vessel frozen, he had eliminated all animal heat and could witness the complete suspension of life. Upon thawing, the rewarmed fish would begin to move, demonstrating physically that life had been maintained all the while.

Unfortunately for Hunter, that failed to happen. According to his account of the experiment, things didn't go so smoothly: "The snow round the carp melted. I put in more snow, which melted also. This was repeated several times, till I grew tired, and I left them covered up to freeze by the joint operation of the mixture and the atmosphere. After having exhausted the whole power of life in the production of heat, they froze; but that life was gone could not be known till we thawed the animals, which was done very gradually. But with their flexibility they did not recover action, so that they were really dead."

His experiment had floundered. Hunter was not, however, dissuaded. In fact, he continued his freezing experiments for decades, documenting them meticulously. In addition to carp, he also tried to freeze and reanimate mice and toads, and various tissues belonging to them. All proved unsuccessful.

What an ongoing bummer! Hunter had failed to support his vitalist conviction that life, at its most fundamental, was a mysterious self-preserving power. Not only that, this failure precluded another possibility that he had been very eagerly hoping for. By freezing and thawing living organisms, he would simultaneously make key inroads into a form of time travel! It seems this may actually have been his primary motivation for the cold experiments, and for keeping them going for so long. Yet, besides making him rich, such discoveries would have changed the course of history.

Till this time I had imagined that it might be possible to prolong life to any period by freezing a person in the frigid zone,
as I thought all action and waste would cease until the body
was thawed. I thought that if a man would give up the last
ten years of his life to this kind of alternate oblivion and action, it might be prolonged to a thousand years; and by getting
himself thawed every hundred years, he might learn what had
happened during his frozen condition. Like other schemers, I
thought I should make my fortune by it; but this experiment
undeceived me.

Simple life, as it were, was not so simple after all. Yet Hunter,
through his many experiments, did make valuable insights into
the preservative, life-saving effects of hypothermia. He revealed
some of the earliest discoveries of the effect of cold on various
tissues, and he explored temperature regulation in a trailblazing
era of scientific experimentation.

◇◇◇◇◇◇

STILL, THE MYSTERY behind the essence of life and the ability to
determine if a being was alive or dead was becoming increasingly
relevant. Aside from its philosophical implications, it had a very
practical one—premature burial. If a body could appear void of
any sign of life but in actuality remain for extended periods in
a state of suspended animation, was it not possible to be buried
alive? And then, in extreme irony, to die a terrible death by suffo-
cating in one's grave?

This concern gave rise to all manner of devices and "safety
coffins" through which it was possible to signal that one was still
alive. While the line between life and death grew increasingly
obscure, the coffins caught on culturally. They incorporated various designs whereby someone buried alive could communicate
with the world six feet above, often by means of a rope tied on

one end to their arm or leg, and on the other end to a bell, flag, or even a firecracker or rocket above ground. These implements, however, were known to procure false positives because natural limb swelling and movement during putrefaction was sometimes enough to engage the alarm. More elaborate designs included ladders and escape hatches. One even included a feeding tube. Ironically, for anyone actually buried alive and able to respond, most safety coffins did not include an air tube. Moreover, someone caught in a live burial situation would likely die of hypothermia if not properly insulated while underground, especially during colder seasons.

Despite his failure with suspended animation, Hunter was becoming a distinguished scientist and surgeon, and his experiments didn't go unnoticed. Through his advocacy of experimentation, he would eventually make valuable discoveries concerning human teeth, bones, inflammation, gunshot wounds, venereal diseases, digestion, and blood. His legacy was secure.

In the 1770s he was approached by doctors William Hawes and Thomas Cogan, the founders of the Royal Humane Society in London, as an expert in suspended animation. From Hunter, they requested a written guideline for reviving victims of apparent drowning, an issue that was increasingly common at that time. They were pioneer advocates of resuscitation, inspired by the Amsterdam Society for the Recovery of Drowned Persons, which had been created to explore apparent death and to discover and promote lifesaving interventions for drowning victims. Equally motivating was a growing number of successful revivals of victims who had been submerged under cold water for extended periods of time. These victims, who initially showed no vital signs before being resuscitated, otherwise may have been taken for dead. (As mentioned in Chapter 2, swimming was not a common activity back then, and relatively few could swim. Drowning was an issue in coastal towns and in areas where seafaring was common.)

Hawes and Cogan proposed that life could exist beyond any overt signs. The number of victims that had successfully been revived—even though they appeared deceased, as evidenced by an absence of breathing and heartbeat, and were cold to the touch—was beginning to suggest either that miracles were becoming the norm or that medical practice could be used to revive the apparently dead. They set out to prove that drowning victims could be resuscitated using new practices that went beyond standard last-ditch efforts, which typically involved bloodletting and tobacco-smoke enemas. In an issue of the Royal Society's journal *Transactions*, they despaired, "Can we reflect on the vast numbers of the human beings, that have been sacrificed in all ages and in all countries, and not feel the utmost remorse, and the most poignant regret?"

In response to their request for advice on reviving drowning victims, Hunter wrote a brief guideline, *Proposals for the Recovery of People Apparently Drowned*, in which he stated, "I shall consider an animal, apparently drowned, as not dead; but that only a suspension of the actions of life has taken place. . . . The suspension will frequently last forever, unless the power of life is roused by some application of art."

In reference to such "art," Hunter advocated using a double-bellows device that could both push air into the lungs and draw air from them in an alternating cycle. *Artificial respiration*, as it became known, was soon advocated for saving drowning victims. Because it proved effective for saving lives, various other techniques for artificial respiration were explored, eventually refining themselves into the CPR technique used today.

## NERVOUS BREAKDOWN

Around the time that Hunter was making inroads into the study of anatomy and advancing surgical methods (while failing to revive frozen animals), other scientific attempts were being directed toward elucidating the mystery of the ill-defined concept of vital capacity. Revolutionary experiments implicated the nervous system as a possible seat for the vital "soul." These discoveries resonate even today, as brain death is now an official criterion for declaring expiration.

The father of modern physiology was the Scottish physician William Cullen, who believed that new physiological research supported the idea that the nervous system potentially held the "essence of life." Specifically, he thought it held the "vital force of sympathy" that animated the human body, coordinated function, and transmitted sensation to target organs. (Indeed, the term "sympathy" is still used in physiological nomenclature; it is associated with the *sympathetic nervous system*, which facilitates involuntary responses, like heightened alertness and elevated heart rate, to potentially threatening events.)

In 1774, Cullen wrote, "Life does not immediately cease upon the cessation of the action of the lungs and heart." Furthermore, "the living state of animals does not consist in that alone, but especially depends upon a certain condition in the nerves and muscular fibres, by which they are sensible and irritable. . . . It is this condition, therefore, which may be properly called the vital principle in animals and as long as this condition subsists," the potential for life exists.

Cullen's speculation on the nervous system did not have any direct relation with cold. His impact on cold science is that he is credited as the inventor of artificial refrigeration. In 1748, as part of his chemistry research, he used an experimental setup involving a partial vacuum to boil diethyl ether. This produced a gas that effectively absorbed surrounding heat, causing ice to form. At the time, he noted that he was able to create ice, but the thought of harnessing the reaction for the potential of cooling and preserving food, let alone any direct medical use, did not follow.

Cold as a means of preserving nervous functioning was, however, noted by the Italian physicist Giovanni Aldini. In a series of impactful, and disturbingly macabre, experiments, he essentially attempted to raise the dead. Aldini was a nephew of Luigi Galvani, originator of the eponymous branch of science known as *galvanism*. At the heart of galvanism was the idea that electrical current could be generated by chemical means and applied to biological tissue. In the 1780s Galvani discovered that dissected frog legs would jerk upon electrical contact. He attributed the effect to electricity inherent in fluid within the tissue—"animal electric fluid"—but this theory was contended by his colleague and rival Allesandro Volta. Volta's invention of an elementary battery—the voltaic pile—showed that the electrical stimulation needed to get the legs twitching could originate outside the body.

Aldini defended Galvani's principles. His explorations of electrically stimulated animal ligaments soon turned into hyped public spectacles as much as they were scientific endeavors. Onlookers watched in awe and fear as Aldini, with the mysterious power of electricity, reactivated the heads of oxen, causing their mouths to open and eyes to twitch as though they had momentarily been restored to life while in a horrific, decapitated form.

In Bologna, in the winter of 1803, Aldini caused an even greater stir when he began to experiment with human cadavers. Using the heads and bodies of executed criminals, he created astounding displays of reanimation, as dubious to some as they were convincing to others. The first experiment involved shocks administered to three decapitated corpses, which activated their headless bodies in a manner that Aldini himself associated with suffering a seizure. Regarding one of the cadavers, he made a note about the cold temperature and the "astonishing result" of the galvanic excitation: "I proceeded to the trunk of the second criminal, which I conceived to be most proper for my purpose. I think it necessary here to observe, that the body had been exposed for about an hour, in an open court, where the temperature was two degrees below zero [Celsius]. The muscles of the fore-arm and the tendinous parts of the metacarpus being laid bare, an arc was established from those muscles to the spinal marrow. In consequence of this arrangement, the fore-arm was raised, to the great astonishment of those who were present."

An even more riveting and "frightening" display was produced by creating a flurry of facial movements simultaneously in two decapitated heads:

> I placed the two heads in a straight line on a table, in such a manner that the sections of the neck were brought into communication merely by the animal fluids. When thus arranged, I formed an arc from the pile to the right ear of one head, and to the left ear of the other, and saw with astonishment the two heads make horrid grimaces; so that the spectators, who had no suspicion of such a result, were actually frightened.

Aldini's ultimate aim was to reactivate a heart and to witness it contract. However, his attempt failed. Still, the hype surrounding his experiments enabled Aldini to embark on a European tour, during which he conducted his showiest and most famous stunts. He wanted to educate the aristocracy on the importance of animal electricity and galvanism. In one experiment, galvanic stimulation induced a violent response when the criminal's body and head were both activated. According to Aldini, "By again applying the arc, according to the method detailed in the 41st experiment, the violence of the contractions was much increased. The trunk was thrown into strong convulsions; the shoulders were elevated in a sensible manner; and the hands were so agitated that they beat against the table which supported the body."

Aldini was most eager to attempt the reactivation of a freshly executed criminal, as all his other subjects had been decapitated. On notably cold January 18, 1802, George Forster (Foster, according to some records), convicted of drowning his wife and infant child in a canal, was hanged at the Newgate gallows. He was twenty-six years old and described as having a "strong and vigorous" stature. After the execution, his corpse remained outside for hours, in the 30°F (−1°C) cold. To start his presentation, Aldini applied electrodes to the head, causing the eyes to open widely and the jaw movements to turn silent screams and facial contortions into various expressions of madness. Aldini then relocated the electrodes to the body and upon closing the circuit apparently caused it to writhe about, legs and arms to move, fists to clench, and head to wag side to side. For the grand finale, Aldini excavated Forster's chest cavity to expose his still heart. Aldini applied the electrodes. Oh, the suspense! The crowd's anxiety, horror, and excitement peaked at the instant when the switch was thrown; the next moment would reveal if Forster could be resurrected.

His heart began to quiver but did not actually contract.

Aldini made notes about the consistently cold temperatures in which his cadaver-subjects were stored for long durations. At the time, however, cold's ability to slow metabolism and preserve biological tissue after a loss of oxygen and nutrient-rich blood flow wasn't fully understood. Although the cold storage of his cadavers may have aided in preserving nervous system functionality, Aldini, perhaps ironically, expressed an element of surprise at the effectiveness of galvanic stimulation *despite* the cold. He surmised that as time progresses,

> moisture performs a conspicuous part in producing contractions; and that it is even of more importance than animal heat. I indeed find that muscular contractions may be obtained after the body has thrown out a great deal of its heat, even when it has cooled for several hours, and when it has been exposed to a temperature below zero [Celsius]; for, if Galvanism be communicated to a body in that state, muscular contractions will be immediately excited; but they soon cease by the privation of animal moisture.

Despite his conclusion about the impossibility of electrocuting executed criminals back to life, Aldini's displays made a lasting impression, in terms of both their spectacle and the insight they provided into the interaction between the nervous system and electricity. In fact, they supported the idea that electricity was a central component of the vital principle. His discoveries were influential for decades, both in and outside the scientific community.

Indeed, they inspired a young author named Mary Shelley, who penned *Frankenstein*, published January 1, 1818. Her story regarding Victor Frankenstein's "monster," who was created from dissected human parts that had been reconnected and galvanically reanimated, weaves themes of warmth and cold throughout the narrative. It also emphasizes the transitions between the two states as a metaphor for suspended animation being neither a living state nor an inanimate one.

Frankenstein's creature has a natural affinity for cold, stating, "I was better fitted by my conformation for the endurance of cold than heat." In part because of this affinity, he attempts to escape the turmoil created during his brief interaction with society by journeying to the northern polar ice cap. He and his creator meet in cold locations, near mountain peaks and in icy caves. Frankenstein pursues the monster into the Arctic but becomes stranded on an iceberg until his rescue by the story's narrator, a ship's captain who is searching for the fabled Northwest Passage, a northern transoceanic route over Canada connecting the Atlantic and Pacific.

The essential role of cold in Mary Shelley's novel has not gone unnoticed. In a comprehensive review of therapeutic aspects of hypothermia, neurosurgery resident Michael Bohl and his collaborators write, "Shelley's novel is revealing not only of her era's infatuation with discovering the principle of life, but also of a growing philosophical understanding of death and life, cold and warmth, and the suspended state of animation that is achieved via hypothermia."

## THE END OF AN ERA?

The prospect that living beings can exist in a frozen state of apparent death, or suspended animation, lives on today, though primarily in the realm of science fiction rather than actual science. Think of movies like *Alien, 2001: A Space Odyssey, Planet of the Apes, Avatar,* and *Star Wars: The Empire Strikes Back,* all of which involve characters who travel vast expanses of time and space while in suspended animation. With the limited methods and technology in place during the eighteenth and nineteenth centuries, however, experimenting with cold was a messy, murky, inconclusive exploit that often created more questions than answers. With neither controlled refrigeration nor thermometers, to name just two modern advances, researchers found it impossible to know the exact degree to which a given specimen was internally frozen. Despite the attempts made by various scientists like Hunter, no conclusive revivals were obtained with animals beyond those only visible under a microscope. Artificial means of suspending animation did not appear to be a fruitful pursuit. In some ways, the outlook for warm-blooded animals—mammals—seemed even bleaker, as they were regarded as less tolerant of cold. Hunterian notions of freezing oneself to travel into the distant future seemed as fantastical as creating a man from a menu of cold anatomic constituents, as in Shelley's famous novel.

Still, despite such grim odds, the science of suspended animation underwent a focused yet notable revival. During the first half of the twentieth century, lesser-known experiments revealed that a variety of insects had the ability to survive freezing during winter. Meanwhile, unsubstantiated claims of fish being able to survive freezing continued, in both popular writing and scientific articles. The main thrust of the revival, however, came rather serendipitously. Bizarrely, it initially involved experiments using a bold new technique that carried such promise that it was immediately tested on mammals: hamsters, rabbits, and even primates.

It all began in the early 1950s in the former Yugoslavia, with Radoslav Andjus, a young physiologist graduate student from Belgrade. It was known by this point that rats were highly unlikely to survive prolonged exposure to temperatures below 59°F (15°C). The logic behind this thermal barrier was that if cooled animals had lost their heartbeat and respiration, reheating their entire body would place too great an oxygen demand on their blood supply for a successful resuscitation. Thus, if rats as fellow mammals were an accurate-enough model for humans, exploring the possibility of life after cooling to 59°F (15°C) was a futile endeavor for investigation.

However, a fire at the university library prevented Andjus from accessing the relevant literature concerning this thermal barrier. So he had no hesitation in entertaining the possibility that revival could be successful after apparent death from hypothermia. He intentionally used rats as a comparative model. Previously, this 59°F (15°C) barrier had been determined by simply rewarming the animals in a heated environment after they had been cooled. Andjus's clever invention was to isolate the heart for rewarming while the rest of the animal remained cold, to place as little oxygen demand as possible on the system during hypothermia.

He cooled the rats in glass jars filled with ice to rectal temperatures below 36°F (2°C) and revived them even after their heartbeat and respiration had ceased for forty to fifty minutes. With this triumph, he established a new principle: even warm-blooded mammals could undergo hypothermia to the point of cardiac and respiratory arrest for extended periods, during which, by all conventional definitions, they were considered dead, only to be resurrected with their internal tissues and organs undamaged.

The next step was to get even cooler—to below freezing. Andjus began a collaboration in England with a young, relatively unknown researcher named Audrey Smith, at the Mill Hill National Institute of Medical Research in London. (Smith is now considered a key originator of the scientific field of cryobiology, the study of biological processes at low temperatures.) Smith and Andjus chilled hamsters in a liquid solution that registered 23°F (−5°C). The animals' deep-body temperatures showed that internal freezing began to occur within a few minutes. In the early experiments, they registered core temperatures of 31°F (−0.6°C) for intervals between sixty and ninety minutes and were demonstrably rigid; they passed the "flop test," meaning their bodies remained stiff when being supported by only head and tail. Upon heart-specific rewarming, they were intubated by having air blown into their lungs with a small straw. Nearly all of them were successfully resuscitated.

Their survival, however, came at a cost: the unfortunate, unethical methods used in the experiments. The unsuspecting rodents were subject to frigid temperatures without being sedated beforehand. After more than an hour of being exposed to potentially lethal cold, their still hearts were targeted for rewarming simply and crudely, by the use of a scalding-hot spoon heated over a Bunsen burner flame. I imagine as well that it must have been an unpleasant, odorous process for Andjus and Smith, with the smell of burned hamster hair and flesh released with each revival, not to mention their awareness of the animals' suffering. Fortunately, this situation was about to change for the better.

In what most would assume to be an unrelated event in London in 1954, a young scientist was combing the streets around Leicester Square for a secondhand market where he could find a very particular piece of electrical equipment. The scientist was James Lovelock, a now-honored independent researcher, environmentalist, and futurist best known for originating the controversial Gaia hypothesis regarding the Earth's systematic interaction between organic and inorganic matter. He was searching for a microwave emitter that he could use to rewarm hamster hearts, as a new collaborator in an early, exciting cryobiology project. In his 2000 book, *Homage to Gaia*, Lovelock reflected on his experience:

> The method Andjus first used to reanimate small animals from just above the freezing point was to apply a piece of hot metal to the animal's chest above its heart. This procedure warmed the heart and started it beating while the rest of the animal was still cold. . . . Audrey drew on Andjus's experience and warmed her frozen hamsters by applying a teaspoon heated in the flame of a Bunsen burner to their chests. This technique worked with some of the frozen animals, but at the cost of badly burned chests. The experimental biologists at Mill Hill were tough and unsentimental about animal suffering. They were not consciously cruel and did try to avoid suffering so long as it did not interfere with the scientific objective of their experiments. This was, I think, the usual attitude of almost all scientists who used animals in the 1950s. I had to be there to monitor the physics and chemistry of the animal as it went through the freezing and re-warming. I soon found that I was made of softer stuff and was repelled by what I thought were cruel experiments.

Lovelock was able to find an old aircraft frequency emitter— a magnetron—and afterward the experiments really took off. Using *diathermy*, the heating of internal tissue with high-frequency electromagnetic current, the researchers were able to resuscitate hamsters that had been supercooled to as low as 22°F (−6°C). In some of these animals, ice crystals could be seen in their internal tissues and blood. What also amazed the trio of Andjus,

Smith, and Lovelock was that most of the subjects remained free of frostbite unless the hamster's paws or ears had been bent before freezing.

The next step was to try the procedure on larger mammals. Hamster reanimation was followed by attempts on rabbits and on small primates—fifteen young-adult Dutch rabbits and six galagos, to be exact. Galagos, also known as bush babies because of their baby-like cries, come from sub-Saharan Africa and are revered for their remarkable jumping ability. The animals were anesthetized and kept in a cool environment before being immersed in baths of 23°F (−5°C) liquid. After some trial and error, followed by improvements involving rectal saline cream and the shaving of insulating belly fur, the researchers were able to cool the rabbits to a temperature of 31°F (−0.6°C). The galagos were also cooled to body temperatures below freezing.

Between two and three hours after cooling began, revival attempts were initiated. Again, the hearts were targeted with microwaves, and artificial respiration was administered. Some rabbits were put in an incubator, others were placed under a warm lamp, and still others were immersed in warm water. After careful rewarming, the first rabbit regained a heartbeat, followed by the next, and then still more. They began to make spontaneous movements; after some time, a few even sat up and moved around.

Within about an hour, however, hopes were severely dashed. The reanimated rabbits all collapsed and died. Same for the galagos. Although their deaths obviously had something to do with cooling, the exact cause—in terms of a specific effect of the cold—remained elusive. The only obvious injury in any of the animals was a ruptured stomach. The experimenters resumed with more animals of the same species. This time they were treated with a bicarbonate before being frozen, in hopes of eliminating any rupturing. After about forty-five minutes of freezing, they were rewarmed with microwaves, and again they regained heartbeats and respiration and began to move about. The early indication seemed excellent. One of the galagos even regained an appetite.

By the twenty-fourth hour, however, disaster. Again, the animals had all expired. No further experimentation was resumed.

These results shut the lid on further scientific investigation into the possibility of live mammals surviving a frozen state. The researchers deduced that it was impossible to temporarily freeze a human being or place one in a state of suspended animation. In the conclusion to a 1958 article in *New Scientist* magazine, Smith wrote, "It seems unlikely that, in the near future, such techniques will be applicable to man."

Rather than a key to prolonged life, it seemed that cold was nothing more than a slow, silent killer.

◇◇◇◇◇◇◇

*After finally surrendering to exhaustion you feel a tidal wave of sleep about to hit you. You succumb to it.*

*In your last moments of semiconsciousness, you open your eyes for a moment and are surprised to find yourself in utter darkness. It's unlike any surrounding you've ever experienced. What could possibly be going on? Have you entered some kind of void where there is absolutely nothing? The space seems infinite. This universe consists entirely of you.*

*When the situation couldn't be more unbelievable, you feel a strange but wonderful presence. Yet you fail to physically perceive anything. You firmly believe that a breathtaking sight is directly behind you. Silently and invisibly, it beckons you to turn your head. When you do, what you see is extraordinary: an incredibly bright but seemingly sourceless light. It seems to extend infinitely.*

*Something like this must go beyond a purpose. It just is. It consumes you until you feel that you're part of it. You don't perceive your own body. You don't even feel as though you have one. Still, you've somehow retained a disembodied awareness.*

*The experience transcends any notion of reality. This is a new real. A new realm.*

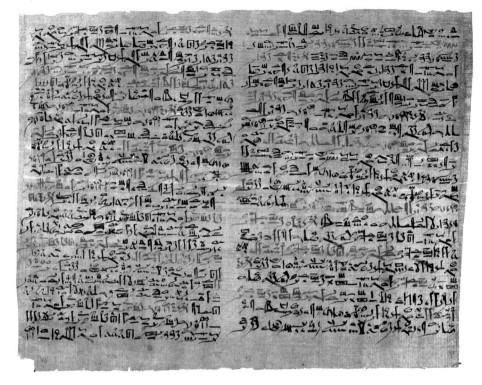

Plates vi & vii of the Edwin Smith Papyrus at the Rare Book Room, New York
Academy of Medicine. *Photo is in the public domain.*

Adolph Northen (1828–1876): Napoleon's retreat from Moscow, 1851.
*Photo is in the public domain.*

*(Left)* An asylum patient receiving a cascade of water as a form of treatment. *Wellcome Library no. 16538i Photo number: V0011765.*

*(Below)* A douche contraption made to target the patient's head. *Wellcome Collection gallery (2018-03-29).*

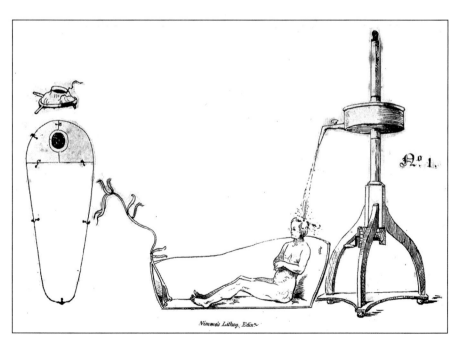

Tales of fish being revived after freezing were an early inspiration for cold science. *Anonymous photographer.*

Behold Gods providence.

Jan: 14. 1651

A woodcut from *A Wonder of Wonders* depicting the hanging of Anne Greene, which she survived.
*Author: W. Burdet, 1651.*

A man recuperating from nearly drowning at a receiving-house of the Royal Humane Society, after resuscitation by W. Hawes and J. C. Lettsom. *Engraving by R. Pollard, 1787.*

An instructional photo showing artificial respiration of an apparently dead victim of drowning. *Wellcome Images.*

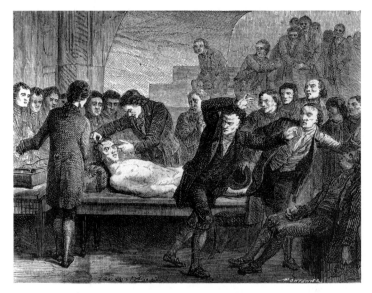

An illustration of a horrified audience witnessing a corpse's face contort upon receiving galvanic stimulation.
*1867, by Louis Figuier, Houghton Library, Harvard University.*

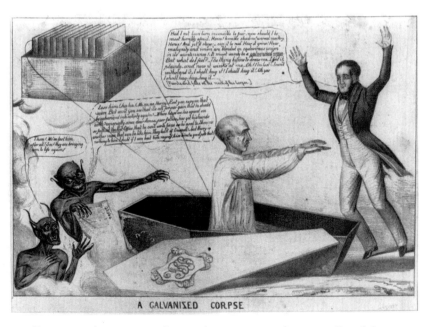

An illustration depicting a galvanised corpse rising from a coffin while demons comment. *H. R. Robinson, 1836.*

Victor Frankenstein horrified at his cold-loving creation.

*Illustration from the frontispiece of the 1831 edition of* Frankenstein.

Cryonicists remove the body of James Bedford from suspension in liquid nitrogen to evaluate his condition.

*Photo courtesy of Alcor Life Extension Foundation.*

One of Dr. Robert White's "living" brains, surgically connected to a blood pump for oxygen and nutrient supply. *Photo courtesy of the Dr. White Archives.*

Dr. Robert White experimenting with a "living brain."
*Photo courtesy of the Dr. White Archives.*

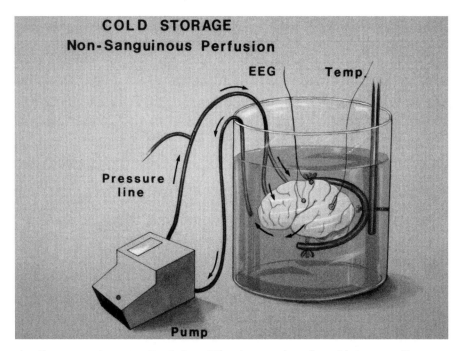

An illustration depicting Dr. Robert White's procedure for cold storage of brains, essentially in vats. *Photo courtesy of the Dr. White Archives.*

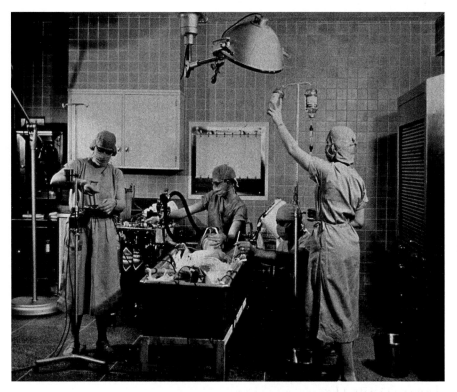

A patient lying in ice before receiving open-heart surgery in 1955.
*Photo by R. Perry.*

# 6

# THE −321°F (−196°C) TIME CAPSULE

## CAN COLD DELIVER YOU TO THE FUTURE?

For around two hundred years, investigators ranging from John Hunter (in the 1700s) to Audrey Smith and James Lovelock (in the 1950s) attempted to suspend the animacy of living beings—from frogs to hamsters—in hopes of reviving them upon thawing. Hunter was determined to discover the essence of life—and also to bag a fortune in creating a means of time travel into the future by freezing oneself to stop the clock on aging and then eventually reanimating. Smith and Lovelock, more practically, aimed to advance medical practices. Regardless, in key respects the process was the same: freeze, wait, reanimate. Success, however, was never achieved. Smith and Lovelock's failure in the 1950s to freeze and revive rabbits and monkeys marked formal science's latest attempt to reanimate frozen mammals. Will it be the last?

With the 1960s came an increase in prosperity in the United States and an optimistic vision toward the future of scientific discovery. Outlandish aspirations like Optiman and the cyborg preceded actual scientific and engineering advancements, such as successfully traveling to the moon and back—achievements that had also seemed like science fiction just years before. Visions of the future went beyond making humans space-proof and greatly expanding the normal lifespan; they came to include the possibility of escaping death completely.

At the time, the list of animals discovered to be capable of self-revival after withstanding freezing core body temperatures was growing. They could survive temperatures that would be fatal if encountered by mammals, including humans. By then it was well-known, for example, that the simple creatures known as tardigrades—one-millimeter-long, barrel-shaped anomalies with four pairs of legs, also known affectionately as "water bears" or "moss piglets"—could survive temperatures near −508°F (−300°C). Increasing in complexity from tardigrades, and probably to John Hunter's dismay had he been alive, researchers discovered that wood frogs could tolerate freezing of up to 60 percent of their entire body while buried in mud during a long winter hibernation. During their period of stasis, they have no heartbeat, don't breathe, and show no sign of brain activity. While frozen, these frogs are dead by all legal and clinical definitions existing today.

What allows them to survive "death"? Basically, the answer involves antifreeze in the form of glucose, a type of sugar. The wood frog carries in its cells high amounts of glucose, which it generates to prevent the destructive effects of ice crystallization, which expands water's volume—think of frozen water bursting a pipe. Truly enigmatic, however, is the frog's ability to literally die and reanimate. In many ways, the current research is similar to John Hunter's quest to understand the nature of life, beyond the beating of a heart and the respiration of lungs. What is the "vital essence"?

Today, our understanding of this frigid frog mystery is the result of cryobiologic inquiry. But aside from inquisitions into the nature of life, cryobiology's primary function lies largely in the lifesaving purpose of organ and tissue preservation. The transplantation of organs, however, can save lives only if the donor organs are viable. Each year, thousands of potential organ recipients die because an organ can't get to them soon enough due to a failure in preservation. The main challenge of this branch of science is to freeze biological tissue for the purpose of preservation, and to do so while maintaining its functionality after transplantation.

These efforts have seen some significant success. Today, many organs, tissues, and cells are preserved at low and even freezing temperatures and then are successfully thawed to resume their function after transplant. A human heart can be stored at around 41°F (5°C) for up to six hours, and blood, semen, and eggs can be stored indefinitely in liquid nitrogen at −320°F (−196°C). At that temperature all cellular processes cease, including those involved in decay. Essentially, liquid nitrogen can freeze time for biological material.

The reproductive cells have a particular resilience after being frozen. Testicular and ovarian tissues can tolerate temperatures fatal to other tissues. Perhaps even you were entirely frozen at some point shortly after your conception. Embryos can be successfully frozen and stored for artificial insemination. In fact, freezing them is routine protocol because inseminations don't always yield a fetus, and multiple attempts are often required.

So, if it's possible to carefully and indefinitely freeze an embryo, what about an entire, fully formed human? After all, in the simplest of terms, we are only composed of more cells, right?

Cryopreservation below freezing seems to only work for cellular preservation and for relatively small amounts of tissue, nothing near an entire body. More than that and the process is likely to generate significant, widespread damage. Sure, you can easily freeze a whole person, but there's no chance they'll live after thawing.

The more relevant question is whether an entire human body can be frozen in a way that minimizes the cellular damage caused by the freezing process.

Some futurists deem this possibility likely. They believe that achieving the necessary level of damage control is a relatively minor detail, and that in the years, maybe centuries, ahead, technology will enable successful reanimation after freezing. Although the notion may sound like far-fetched science fiction, current medical advances have not only enabled the successful freezing and thawing of cells, tissues, and organs; they have procured means of reviving people from hypothermic states in which, by all conventions, they would be declared dead, with no heartbeat and no measurable brain activity. Remember Anna Bågenholm, the skier? What defines death has already been obscured by medical technology. Similar to the wood frog, can a hypothermic human with no vital signs always be considered dead?

That said, none of these futurists have volunteered to be frozen while *alive*, in the manner akin to animals that are capable of surviving freezing core body temperatures. Rather, it seems that such optimism begins with the idea of preservation *after* death. A person would have to rely on the invention of technology that could successfully recover them from their frozen state, and could also cure what killed them in the first place. Ideally, such technology would essentially restore youth. After all, someone who dies in advanced age after suffering the ravages of numerous ailments would likely prefer not to be restored to their preexisting condition.

## ETTINGER'S CHILLING CONCEPTION

One such futurist was Robert Ettinger, the late originator of cryonics, the preservation and freezing of human corpses with the speculative hope of revival in the future. Ettinger was born in 1918 in Atlantic City, New Jersey. A fan of science fiction as a youth, he found one story particularly thought provoking: *The Jameson Satellite*, written by Neil R. Jones and published in *Amazing Stories* magazine in 1931. In the story, Professor Jameson arranges for his corpse to be launched into space in hopes of preserving it as it orbits Earth indefinitely at temperatures approaching absolute zero (−459.67°F or −273.15°C). After no less than forty million years, when the human race is extinct, Jameson is discovered by an advanced race of aliens. Their brains are similar to those of humans, but they have transcended organic bodies for mechanical ones, which allow them to replace broken parts when needed, thereby living indefinitely.

Later in life, Ettinger reflected on how this story inspired him, writing, "It was instantly obvious to me that the author had missed the main point of his own idea! If immortality is achievable through the ministrations of advanced aliens through repairing a human corpse, then why should not everyone be frozen to await later rescue by *our own people?*"

And thus, cryonics was conceived.

As a young adult, he served as an infantryman during World War II. Severely wounded in battle while in Germany, he received the Purple Heart award. His wounds may have been fatal had it not been for cutting-edge bone-grafting techniques. The procedure was considered experimental at the time, and Ettinger was counted as a success story. Undergoing successful surgery thanks to advanced medical technology and receiving the Purple Heart inspired Ettinger to seriously consider the idea of preservation after death. The hope was that future medical advancements could revive and repair a human's body, returning him or her to a comfortable and indefinite state of being. Ettinger assumed that someone, perhaps a scientist, would advocate the idea, turning it into an area of research without the need for any active involvement on his part.

He went on to study mathematics and physics and became a professor at Wayne State University and at Highland Park Community College. By 1960, at age forty-two, he began to consider his own mortality. As far as he knew, nobody else had realized his idea for escaping death. If he wanted to pursue the possibility of saving his own life, he'd have to act right away. And so he wrote down his conception of frozen stasis and reversible death and sent the paper out to two hundred people—recipients whom he'd carefully selected from a list of movers and shakers culled from *Who's Who in America*. After his attempt failed to conjure any significant support, he wrote that, to his astonishment, "a great many people have to be *coaxed* into admitting that life is better than death, healthy is better than sick, smart is better than stupid, and immortality might be worth the trouble!"

His next step caused the stir he was looking for. He wrote a book. After a preliminary copy was promoted by Isaac Asimov, *The Prospect of Immortality* sold exceedingly well. Ettinger became an overnight celebrity. Reviewers considered the book profound. It was covered in the *New York Times*, *Newsweek*, *Paris Match*, *Der Spiegel*, and (appropriately enough) *Time*. He appeared on radio and television talk shows and was interviewed by David Frost, Johnny Carson, and Steve Allen.

Later, reflecting on why it had taken him so long to write his book, Ettinger said, "For the simple reason that I had, and have, no credentials worth mentioning, being only a (now retired) teacher of college physics and math. It is precisely this that prevented me, for so long, from doing more: I knew I carried no weight, had no formal qualifications, and was not suited for a leadership role. But as the years passed and no one better came forward, I finally had to write."

Despite his self-effacement, Ettinger was in fact a fantastic candidate to advance a far-out idea. An academically credible war veteran, he was an articulate spokesperson who fit the creative and optimistic spirit of the 1960s.

Over the next couple of years, various cryonics-centered organizations were born. The first, called Life Extension Society, was founded in 1964. (It wasn't until 1965, however, that an industrial engineer and cryonicist in New York by the name of Karl Werner invented the term "cryonics" by fusing *cryo-* and *bionics*.)

Other groups formed in California and Michigan, where Ettinger was president of the Cryonics Society of Michigan. Yet something crucial was missing: the first patient!

Evan Cooper, who founded the Life Extension Society after reading Ettinger's book, made extensive efforts to promote cryonics to both scientists and the public at large. Dismayed by the lack of appeal generated after two years, he wrote, "Are we shouting in the abyss? How could 110 million go to their deaths without one, at least trying for a life in the future via freezing? Where is the individualism, scientific curiosity, and even eccentricity we hear so much about?" The Life Extension Society even offered to freeze the first volunteer free of charge.

The excitement, motivation, and eagerness displayed by the advocates of cryonics were aimed at catapulting the theory into practice in hopes that it would spur the world into realizing that immortality was within the limits of science. Choosing cryonics seemed so obvious and logical. If you like life today, why not take the steps toward a world where each new day brings with it the possibility of another? Surely any chance is better than a guarantee of rotting or turning to dust if there's no afterlife.

Although many cryonics promoters held this attitude, they advocated the slow, gradual process of science—experimental testing and formal research conducted according to a rigorous, skeptical standard—to achieve their aims.

In 1967, a golden opportunity presented itself. A series of events that would generate worldwide news about cryonics was about to unfold. The first volunteer had arrived. A patient diagnosed with terminal cancer expressed interest in being frozen. Of course, the actual science of cryonics as a practice was still in its infancy.

The stage was set with a call between Robert Nelson, an ex–TV repairman who was president of the Life Extension Society, and Robert Ettinger. Nelson recalled, "Well, I called Robert Ettinger that night, and I told him what had happened. And he said, oh my god, this is the biggest thing that's happened in the cryonics program. And so Ettinger said, we need to go ahead and do it. And I said, but we'll lose the scientific advisory council. He said, maybe not all of them. And if we do, we'll get them again. He said, there's nothing that will push the program of cryonics forward [faster] than the freezing of the first man."

## JAMES "COOL DUDE" BEDFORD

James Bedford, born in 1893, was a psychology professor at the University of California who specialized in occupational counseling. Bedford loved photography and traveling with his wife, Ruby. They had five children. Bedford suffered from kidney cancer that had metastasized into his lungs, becoming untreatable. He was an ardent supporter of cryonics, having left a sum of $100,000 in his will to further the research. The Life Extension Society had offered to cover the cost for the first volunteer to be frozen, but Bedford wouldn't have it that way and instead paid for his own preservation.

He thought that his reanimation would likely be unsuccessful; his main motivation was spurring interest in cryonics. Bedford believed that his involvement as the first cryonaut would help advance funding and research. Hopefully, by the time his children passed on, the methods would be advanced enough to give them a second chance at life.

The freezing of his body was notable, both for the efficiency with which the body was initially processed, and also for the extended delay that transpired before it was stored for the long term in a liquid-nitrogen capsule. The procedure itself, however, was unprofessional, DIY, hacker science, crude and amateur, and followed no official standard.

The series of events began on January 12, 1967. The conditions for getting Bedford frozen quickly after his death, before too much tissue damage could occur, seemed ideal. Bedford was housed at a nursing home, better for these purposes than a hospital, where the red tape of legal requirements and administrative procedures would have interfered with the preparation of the corpse for cryonic stasis. On call to prepare and freeze Bedford's body were members and affiliates of the Life Extension Society: Robert Nelson, the organization's president; Robert Prehoda, an author and cryobiology researcher (who had recently become skeptical of cryonics but helped nonetheless); and Dante Brunol, a physician and biophysicist. Bedford's physician, Renault Able, loyal to his patient's wishes, also assisted.

After Bedford took his last breath, the nursing home staff declared death. To begin the freezing process they called Nelson and his colleagues, who apparently were surprised by the timing of the death. In his book, *Freezing People Is (Not) Easy*, Nelson describes what happened:

Robert [Prehoda] and I had just picked up the necessary chemicals and were back in the office when Dr. Able called. "I just pronounced him dead," he said. I was caught off guard and didn't know what he was talking about at first. "Who?"

"Bedford! Who did you think?"

He sounded more than a little anxious. "Did you procure the DMSO?" [Dimethyl sulfoxide is a chemical used in cell banking to prevent the formation of damaging ice crystals in biological tissue.]

"Shit," I said, feeling disoriented. "I thought he had a couple days left."

"Patients tend not to wait for permission. Did you get the goddamned DMSO?"

"Yes. We just got back."

"Good. Please hurry. The scant amount of ice I have is starting to melt. You need to pick up more. He's on the heart compressor, but nothing else can happen until you arrive with those chemicals."

"We're on our way."

The nurses needed to chill Bedford's body with ice until the team could get to the nursing home. Unprepared for this weird task, they ran up and down the block requesting ice from neighbors, gathering a tray here and there along with the odd freezer pack. One can only imagine their replies when asked by these neighbors what exactly the ice was for.

When the team finally arrived, Bedford's corpse was perfused with DMSO, and he was given heparin, an anticoagulant. (Although once thought to be useful for long-term cryogenics, this compound is today regarded as relatively crude.) It is doubtful that these measures, although undertaken with due speed, prevented significant damage to Bedford's brain.

By 3:00 a.m., the process was complete, and Bedford's body was ready for the "big wait" in liquid nitrogen. Although tired, the team felt triumphant. Brunol and Able left for home while Nelson and Prehoda remained at the nursing home. In the rush to prepare the body, they had forgotten a very important detail: the cryogenic freezing capsule that would hold Bedford in liquid nitrogen was still under construction. It wasn't scheduled to be ready for at least two weeks.

At this point, things became darkly comical. Bedford was placed temporarily in a container made of wood and polystyrene and surrounded with dry ice. He was essentially packed into a large Styrofoam picnic cooler. They took his body to Prehoda's house. It seemed to be their only option—at least until they could think of something better. Prehoda, however, knew that if his wife, Claudette, caught on that a corpse lay in their garage, all hell would break loose.

In the dead of night, they used Nelson's truck to transport Bedford to Prehoda's home. Prehoda opened the garage and parked his station wagon inside. He lowered the back seats, and as quietly as they could, the two men hoisted Bedford, in his container, out of the truck and into the back of the wagon. They shut the vehicle's door and closed the garage.

Before calling it done, they decided to give Bedford one last check. After all the trouble they had gone to, it would certainly have been a pity if his body had been damaged. They opened the lid. He appeared to be in good shape (relatively speaking, of course)—except a ten-pound block of dry ice was sitting directly on his face. It had been there since before he was completely frozen and had squished his nose, which was now set to remain awkwardly bent for potentially centuries.

They closed the lid again.

One can imagine Prehoda's predicament if a neighbor, out for a walk while unable to sleep, had been curious enough to ask what they were up to. It was probably the strangest night either of the two had spent.

An account given by another Life Extension Society member, who remains anonymous, reports on the observations of an unidentified witness:

> Eventually Prehoda's wife found out about the body in the station wagon in the garage and . . . she got pretty hysterical. As we understand it the windows of the station wagon were soaped so no one could see in and the wagon was moved. . . .
>
> Our observer gave up describing the scene in detail at that point saying it could only be described as hysterical and chaotic. He said that if he had [had] a camera it would have made the movie of the year.

According to Nelson's account, when Claudette found out about the body, she gave Prehoda six hours to remove it from their property or she would call the police.

They were left to call friends and ask if they could take the body to their house. Thankfully, Nelson had a couple of hippie friends who accepted the request. "What the hell; that's what friends are for!" Nelson recalls them saying. Good friends. So Bedford's container was taken for another ride in Nelson's truck along a winding, bumpy dirt road to a home in Topanga Canyon. Bedford remained there until his cryogenic capsule was finally ready. His containment therein, during which he was immersed in a bath of liquid nitrogen, went successfully.

The event of Bedford's freezing drew headlines in newspapers across the country. Nelson and Ettinger appeared on news programs and talk shows.

And now that the first freezing had actually taken place, what of the Life Extension Society's scientific advisory council?

It disintegrated instantly. A member called Ettinger and said that it was all over. The council members would no longer take part in anything having to do with cryonics, and they requested that their involvement be kept confidential.

Bedford's crude, haphazard, but successful freezing triggered shockwaves that largely dissociated cryonics from the scientific fields of cryogenics and cryobiology. These areas of inquiry involve the formal study of materials and biological processes at low temperatures as conducted by academics and paid professionals in established institutions and private industries. Well into the 1970s, cryonics became further entangled in slapdash proceedings that propelled it into a world of lawsuits, animosity, and, worst of all, thawed, rotting corpses.

## THE CHATSWORTH DISASTER

Although Bedford's freezing generated considerable publicity, it failed to produce any significant funding. The society was broken. Nelson needed a plan. He needed enthusiastic volunteers who could offer their services for free or at least at a discounted rate. Specifically, he needed a mortician who could oversee cryo-preservations and maintain bodies at their facility. Surely this would prevent any future freezings from being as haphazard as Bedford's. After receiving letters of interest from an astounding 147 mortuaries, he chose a man named Joe Klockgether.

Around the same time, one of the members of the Life Extension Society died. Her name was Marie Phelps-Sweet, and she had wanted to be preserved. Her death, however, came before Klockgether could offer his services. Again, Nelson was caught off guard. When he showed up at the morgue, he found Marie wearing a medical bracelet stating that she had requested to be frozen.

Legally, the bracelet wasn't good enough to grant Nelson the right to transport her away. Yet he and his colleagues desperately needed to perfuse her body with antifreeze and anticoagulant. Nelson, along with two student morticians, sneaked her out of the morgue and carried her to the only place they could come up with to perform the necessary surgery: the Cryonics Society of California (CSC) office, of course. At that time, Nelson was president of the CSC. He had decided that the perfusion would take place in his office.

Nelson later recalled, "I was a nervous wreck because I'm thinking, I don't know how many violations I'm committing here. For example, a dead body legally can only be moved by a mortician. And then I had no idea if I was committing any violations by having the body up in our offices, and putting her on ice there, and then carrying her down the stairs. It was all just really peculiar."

They ended up preparing her body on top of two desks placed side by side.

After she was put on dry ice, it became apparent, again, that there was no place to store her until a liquid-nitrogen container could be purchased and readied. Phelps-Sweet's body was sent for indefinite storage on dry ice at Klockgether's mortuary.

Within months another death within the society occurred. This time it was Helen Kline, a woman who Nelson claimed had been the one to introduce him to cryonics in the first place. Like Phelps-Sweet, Kline had no money set aside for the procedure and so was frozen pro bono. Hers became the second body stored indefinitely at the mortuary, both housed in picnic cooler–like boxes.

It should be noted that Klockgether's "facility" was actually just his home garage, and maintaining the bodies there was an expensive, effortful job. Nelson claimed he had to provide hundreds of pounds of dry ice every week, driving a couple of hundred miles each time. Klockgether described the tense situation:

> I was anxious to get them out of here: Bob [Nelson], come on, I got to use my garage. I got things I want to do. I don't want to keep doing this here. And I don't want to play around with the health department.
>
> See, there's a term, *temporary* storage. They don't really clarify what temporary means, but you or I know temporary doesn't mean forever. Temporary—something should be down on the road. You should have some kind of a date.

The two bodies remained in Klockgether's garage for the greater part of a year.

Meanwhile, society members, many of whom were old, were "dropping like flies," according to Nelson. In the summer of 1968 a third member, Russ Stanley, suffered a heart attack. Stanley was one of the society's most enthusiastic members. He liked to call Nelson at any hour during the day to talk, sometimes for up to two hours, about the latest in cryonics news. Medics were unable to revive him. After Stanley's preparation, he, too, was placed in an oversized Styrofoam picnic cooler.

Fortunately, before his death, Stanley had donated enough money to the Cryonics Society—between $5,000 and $10,000, depending on whom you ask—to create a proper storage facility. It wasn't until 1970, however, that his vision became a physical reality. It started rather modestly, as a vault in a cemetery in Chatsworth, a neighborhood in Los Angeles. Small, yes, but better than Klockgether's garage.

Still, there were no containment capsules for storing the bodies in liquid nitrogen. This was a concern because although they were stored in dry ice at around −112°F (−80°C), putrefaction of some cell types can occur after just one month at that temperature. The situation was so dire that Nelson resorted to disreputable actions that ultimately took cryonics down a steep, decades-long nosedive from which its reputation has never fully recovered. While the bodies slowly rotted, Nelson appeared on talk shows describing the patients as being stored in futuristic pods similar to those depicted in the Stanley Kubrick film *2001: A Space Odyssey*. He even kept a fraudulent picture to display as "evidence" of his claim.

In addition, Nelson had to avoid Klockgether, who wanted the bodies out of his garage, because he in turn was avoiding the health authorities, who would also have wanted the bodies out of the garage if they had discovered them.

Seemingly against all logic, the CSC agreed to take on another patient. This one, however, was already stored in liquid nitrogen, in a proper containment capsule. The patient, Louis Nisco, had been preserved by a new outfit called Cryo-Care. After the company president, an ex-wigmaker named Ed Hope, realized that cryonics would probably never turn a profit, he left and the business fell apart. Nelson acquired Nisco's body after offering his daughter, Marie Bowers, a lower rate than Cryo-Care's recurring fee for upkeep of the capsule.

Now, with Nisco at Chatsworth, there were three bodies on ice and one in a containment unit. Because replenishing the dry ice was costly and tiresome, Nelson got an idea—a very good, very *bad* idea. He looked at Nisco's pod and made some measurements.

As you've probably guessed, his idea was to put all four bodies in the pod.

Nelson and Klockgether opened the capsule, which was clearly designed for only one cryonaut. They removed Nisco's body along with a device meant to provide support. Nelson has described the ensuing task of arranging all four bodies in the unit as "putting together a Chinese puzzle." Some were put in head first, others feet first. The game of human Tetris apparently took most of the night.

A very agreeable and probably well-paid welder was called in to reseal the capsule, a task requiring hours more work. Substantial detrimental rewarming most likely had occurred. The welder called the experience one of the worst of his life, claiming that he could smell burned flesh and hair as he welded the can shut.

Bowers was never informed that her father would have intimate visitors with no plans of leaving.

Still, Nelson's problems were not solved: "We had to keep a pump, an electronic pump, pulling the vacuum 24 hours a day, seven days a week. At Chatsworth, the temperatures got up to over 100, 110 sometimes. And that was death to these vacuum pumps. They couldn't take that heat. The pumps would burn out, and needed to be replaced. Then it just got worse and worse and worse. I was there, I would say, virtually every day." What's more, just like the dry ice, the liquid nitrogen needed frequent topping up because the capsule's functionality had been reduced when it was tampered with.

For months, Nelson continued his exaggerated, fraudulent promotion of cryonics while barely maintaining the storage facility in a manner suitable for even a B horror film. Then, in 1971, history would repeat itself.

This time he was approached by a man from Quebec, Canada, named Guy de la Poterie. Guy's seven-year-old daughter, Genevieve, was dying of a rare, untreatable kidney cancer. Nelson found Genevieve endearing. He became friendly with the family and even organized a trip to Disneyland for them.

Unfortunately, Genevieve lost her life before her father could secure the money needed for her cryonic preservation. Nelson was present when she died. Ignoring his and the family's finances, he put her body on ice and arranged for it to be perfused with a cryopreservant. She was laid to rest in the same sort of limbo as the others had been, on dry ice without a liquid-nitrogen dewar. Basically she was in a human-size thermos.

Unbelievably, Nelson acquired still *another* body—that of a woman named Mildred Harris, for whose processing he received a sum of $10,000. Yet again, no long-term containment unit existed for her preservation.

Soon thereafter he obtained another corpse that, like Nisco's, was already frozen in a proper capsule. It was the body of Steve Mandell, the first patient of the newly formed Cryonics Society of New York. Again, Nelson got the same good, bad idea. He opened Mandell's capsule and arranged Genevieve de la Poterie's and Mildred Harris's bodies inside, unbeknownst to Pauline Mandell, Steve's mother.

Like the first one, this tank needed frequent top-ups of liquid nitrogen. Thus, at this point, Nelson was responsible for two failing cryopreservation tanks full of bodies. By 1972, the first tank had become too much of a burden, in terms of both time and expense. Nelson's money and patience ran out, and he simply let it fail with the bodies inside. Sordidly, he kept that fact to himself while continuing to maintain the second tank.

In 1974, while Nelson was on vacation, he paid a groundskeeper at the cemetery to take over the capsule's maintenance. During his absence, the tank's vacuum broke and the tank failed. Nelson recounts his first visit to the vault after his return home:

I came back, drove up to the vault, looked at the capsule. There's a nozzle that comes out of the capsule that has steam, visible, because the liquid nitrogen is evaporating away. And when I drove up and I looked, that steam wasn't there. So I just didn't want to acknowledge what that meant. But the test was to go and touch that pipe, and if it was cold, then there was some hope. That meant that it was still cold inside.

And then, going through my mind, what if it's hot? What if those bodies have decomposed? So I walk up to the capsule. I put my finger on it. And it was like touching a hot frying pan.

Nelson's worst fear had come true. There was not a drop of liquid nitrogen left in the tank. How long the bodies had been decomposing was anyone's guess.

According to media reports, when an attorney named Michael Worthington was tipped off by a reporter who had caught wind of the situation at the cemetery, he went to the Chatsworth vault, accompanied by Mildred Harris's son Terry and a small news team. Terry described the event:

Well, the door in the facility was made of steel. And it was then chained and padlocked closed. The chain was rusty, and there was grass growing around that door where, before, it wasn't. And our attorney brought a pair of bolt cutters, and removed that lock and chain, and slid the door back. And we went down, and you could just see that there was a piece of equipment here and there, and the capsule lid open.

And it was unbearable, just unbearable. And I was just—I was just numb. Just numb. Well, I couldn't look inside that capsule, but I just backed away when I realized that there were just remains inside.

One observer said that the decomposing bodies had "sludged down into what I can best describe as a kind of a black goo." *Valley News* reporter David Walker wrote, "The stench near the crypt is disarming. It strips away all defenses and spins the stomach into a thousand dizzying somersaults."

Even more unfortunate is that these disasters weren't the only ones. In fact, failure that resulted in thawing seemed to be the norm, rather than an anomaly. Out of a total of seventeen freezings that were performed before 1974 by various cryonics groups, the only body that remains frozen today is that of James Bedford.

Sign-ups dropped over the next decade. If Nelson's account of cautioning Ettinger against being hasty to freeze the first patient without backing from academics is true, he had ironically played an essential role in destroying the credibility of cryonics as a scientific field, ensuring that professional researchers wouldn't go near it for years.

Dizzying reports detail the legal action taken against Nelson and Klockgether after the nine decomposed bodies were discovered. The lawsuits totaled between $800,000 and $1 million (depending on the source). Apparently Klockgether persuaded his insurance carrier to pay his portion. Nelson fought the claims, asserting, "I haven't done anything criminal, anything wrong other than a lot of bad decisions," and that he'd "never promised anything." He said, "They were told they would be frozen for a period of time. Five minutes is a period of time." Due to procedural irregularities, he negotiated the judgment down to $18,000, which he paid in part by selling his car.

## A NEW BEGINNING

By the last half of the eighties, Nelson's disastrous impact on cryonics had somewhat waned. A new technology was boosting the optimism of cryonicists and increasing the appeal of cryonics with the public. It was *nanotechnology*: the study, creation, and application of objects on a nanoscale—that is, measuring between 1 nanometer (one-billionth of a meter) and 100 nanometers.

When cryonicists became aware that it might be possible to manipulate, rearrange, and organize matter on minute scales, right down to the atomic level, it didn't take long for them to realize that nanotechnology could, in theory, repair frozen human tissue upon thawing. In fact, the connection between cryonics and nanotechnology can largely be traced to a single source, a 1986 book titled *Engines of Creation* by engineer Eric Drexler. Envisioned within its pages is a world where nanoscale machines—"universal assemblers," as they are called—build matter, creating objects that fulfill a variety of tasks. He imagined, for instance, that such machines could store the entire contents of the Library of Congress within the physical space of a sugar cube. They could also carry out medical procedures—anything from clearing capillaries to, well, repairing dead people. In fact, he included an entire chapter devoted to nanotechnology's potential for cell repair, with cryonics in mind.

While *Engines of Creation* was giving new life to cryonics, another brush between cryonics and the law took place. In 1987, Saul Kent, one of the founders of the New York Society for Cryonics, had arranged for his mother, Dora, to be cryopreserved by a company called Alcor—currently the largest existing cryonics firm. Dora Kent was Alcor's eighth cryopreservation. As a "neuro-patient" who had Alzheimer's, she was not to be reanimated in her compromised, elderly form. She died at the Alcor facility, and workers removed her head to process and freeze it. The rest of her body was delivered to a coroner, who issued a death certificate stating that she had died of pneumonia.

However, the coroner's analysis also revealed the presence of a barbiturate in Kent's blood. According to Alcor, it had been administered after her death to slow brain metabolism, thereby reducing damage before cryopreservation. The coroner was unable to determine if the drug had indeed been administered postmortem. In fact, it would have been beneficial for the cryopreservation process for Kent to have received the barbiturate *before* her death, to minimize cell damage. The coroner became suspicious that she might have been killed for the sake of an optimal preservation. He changed the cause of death to homicide and demanded her head, which by then had been frozen in liquid nitrogen, for further analysis along with other bodies at Alcor.

When the police went to the Alcor facility, they received no cooperation from the staff and were unable to obtain the head and bodies. Later, a SWAT team raided and ransacked the facility, seizing computers and records. The staff were taken away in handcuffs and arrested. Kent's head was never found—it had been moved to an undisclosed location.

What at first seemed like another episode of cryonics malfeasance ended, surprisingly, in Alcor's favor. Not only were the Alcor employees and executives declared innocent; a restraining order was granted against the coroner. Moreover, Alcor successfully sued the county for false arrest and illegal seizure, winning a $90,000 settlement. The case received wide media coverage, ultimately yielding enough positive exposure to generate greater public acceptance of cryonics—and a sharp increase in sign-ups for the process.

At the same time, significant progress was being made in cryopreservation science. The preparation process was being improved, and the field began to receive public support from mainstream scientists. These advancements were largely to due the efforts of a man called Mike Darwin, a young cryopreservation wiz who was known for his technical skill. He'd acquired his moniker because of his enthusiasm for biology. Other significant contributions were made by the scientist Jerry Leaf, a Vietnam veteran who studied at Cerritos College, in California, and then became a cardiothoracic researcher at the UCLA School of Medicine. Leaf authored upward of twenty research papers before founding his own cryoresearch facility, which was contracted by Alcor to provide cryopreservation services. It was cryonics meets academic science on a formal level.

In the mid-1980s, Leaf and Darwin conducted experiments with dogs at Leaf's facility. With the aid of a blood substitute they had engineered, they were able to revive the dogs without neurological damage after the animals had been exposed to deep hypothermic conditions in which their bodies approached the freezing point and were kept there for hours. Yet cryonics remained ostracized by the cryobiology community, and their research was not accepted for publication in peer-reviewed academic journals.

But that wasn't the end of it. In 1987, inspired by Leaf and Darwin's efforts, Paul Segall, an associate professor of physiology at the University of California at Berkeley, conducted similar research. He and his colleagues were also able to revive dogs after deep hypothermia. This time the research was published, in the journal *Biomedical Advances in Aging*. Segall, a member of the American Cryonics Society, was among the first cryobiologists to openly support cryonics, and his article generated considerable positive publicity for the process.

## CRYONICS MATURES

Since then, the construction of patient-housing dewars has much improved, and they now have little likelihood of breaking down. Moreover, a better financial model exists; a patient can designate their chosen cryonics service provider as a beneficiary of their life insurance policy, and money for their future revival is often placed in a trust fund. No more pro bono freezings.

Although scientists like Darwin and Leaf have made significant improvements, there's still a long way to go. Yet, cryonics organizations are more open about the considerable risk of cell damage due to freezing and the unknown likelihood of revival. Alcor's website, although many may argue that it is scientifically misleading, confirms that disclosure is a priority of the company. The site provides information about biological damage resulting from freezing, details the shady history of cryonics, including the tragic events covered in this chapter, and clearly states that future revival is not in the contract.

Other significant advances have been made in the methods and materials used in the cryopreservation process. A common myth holds that cryopreservation simply involves one's recently dead body being dumped in a tank of liquid nitrogen, which is sealed shut and indefinitely stored in a vault. Granted, the early preservations were relatively crude and likely resulted in considerable damage to the corpses. By contrast, in terms of the body's preparation, the current process is remarkably complex. It involves a host of intricate measures to prevent cell damage caused by the expansion of water within and between cells when frozen, techniques that have been developed through rigorous science over decades.

The most significant of the new processes is *vitrification*. Technically, vitrification is defined by a fluid's increase in viscosity to the point where its atoms are connected to a greater degree and it no longer has the capability of flowing. After receiving injections of vitrifying solution, a cryopreserved body is cooled to around −184°F (−120°C). At that temperature, known as *glass transition*, the body reaches a glass-like state in which it is neither a solid nor a liquid. Only then is it finally immersed in liquid nitrogen, which slows molecular activity so that no biological processes can occur. While the body is cooled, it is carefully monitored for fracturing.

Currently, an impressive total of nine cryonics service providers operate throughout North America, Europe, Asia, and Australia. Even more impressive is the number of cryopreserved patients, now over four hundred. Prospective candidates number in the thousands.

Despite the increase in hopeful supporters and the improvements in technique, cryonics is still widely considered a pseudoscience, and the academically based Society for Cryobiology still maintains a distance from it. Its statement on cryonics (currently on its website) reads as follows:

The Society recognizes and respects the freedom of individuals to hold and express their own opinions and to act, within

175

lawful limits, according to their beliefs. Preferences regarding disposition of postmortem human bodies or brains are clearly a matter of personal choice and, therefore, inappropriate subjects of Society policy. The Society does, however, take the position that the knowledge necessary for the revival of live or dead whole mammals following cryopreservation does not currently exist and can come only from conscientious and patient research in cryobiology and medicine. In short, the act of preserving a body, head or brain after clinical death and storing it indefinitely on the chance that some future generation may restore it to life is an act of speculation or hope, not science, and as such is outside the purview of the Society for Cryobiology.

It's a surprising statement. At face value, it seems hardly discrepant with the views of cryonics supporters. It appears to make two main points. The first is that revival after cryopreservation is not currently possible and will require painstaking scientific research if it is to become a reality. But cryonics advocates most likely are not praying for the discovery of something truly magical; nothing about cryonics involves a supernatural defiance of the known laws of physics. Of course it will require science.

The second point contends that because revival is not currently possible, the practice of cryonics is based on hope, not science. The implications here are similarly confusing. Any cryonics supporter probably knows that they have no choice but to *hope* for the best. A 2018 survey of cryonics supporters, conducted by a Purdue University researcher, indicated considerable variability in how members envisioned their potential revival. They varied in terms of whether they thought their own brains were necessary for their revival, or whether they believed that their unique self—that is, their memory and identity—could be "uploaded" onto artificial, inorganic hardware.

Additionally confusing about the statement is that if the revival of a preserved human body—cryonics in its purest definition—can be made possible only by "conscientious and patient research in cryobiology and medicine," then it sounds as though it can be made possible only by science, to the extent that cryobiology and medicine are scientific. That contradicts the last part of the statement, which dissociates cryonics from science: "the act of preserving a body, head or brain after clinical death and storing it indefinitely on the chance that some future generation may restore it to life is an act of speculation or hope, not science."

Whatever the current state of the relationship between cryobiology and cryonics, formal scientific research with the aim of carefully preserving tissue, organs, and brains is growing in strides. For example, the UK Cryonics and Cryopreservation Research Network consists of twelve established academic researchers at various universities. Also increasing are the number of research facilities investigating various aspects of cryopreservation that overtly state an aim to advance methods of *human* preservation. There's also the Brain Preservation Foundation (BPF), whose mission statement is "to promote validated scientific research and technical services development in the field of whole brain preservation for long-term static storage."

Since the creation of the BPF in 2010, it has offered two cash prizes totaling $100,000, accrued through donations, to anyone who can demonstrate a long-term method of brain storage capable of preserving individual synaptic connections between neurons for at least one hundred years. With money like that up for grabs, have any of the cryonics service providers entered to win it? Surprisingly, not one. No cryonics service organization has entered a BPF competition. However, the efficacy of a cryoprotectant used by Alcor, called M22, was assessed by a leading cryopreservation organization, 21st Century Medicine (21CM).

The result? Well, it's complicated. Researchers used M22 in a brain-preservation process, but they were unable to analyze the preserved brain with electron microscopy—which is needed to examine cells at a sufficiently tiny scale—because the M22 had caused the brain to shrink to around half its regular size, rendering it unusable for the necessary examination. Although the 21CM scientists hypothesized that the neural connective structure may have remained intact within the brain's misshapen form, they had no way to test their hunch. Still, M22 has been shown to preserve neural connections in tissue extracted from an area of the brain called the hippocampus, which plays a central role in memory.

That said, considerable evidence suggests that M22's effectiveness in preserving the hippocampal region is unlikely to be duplicated in other regions of the brain with different structures and cell types. Moreover, plenty of evidence indicates that M22, which is toxic, may cause substantial damage to brain tissue in the relatively lengthy time between injection of the substance and whole-body freezing. Indeed, cryoprotectant toxicity remains the most significant obstacle to successful cryogenic freezing.

Although most cryonicists believe that cryonics affords only a small chance of successful reanimation, what matters to them is that, unlike the case with conventional burial or cremation, the probability it offers of being brought back to life is greater than zero.

◇◇◇◇◇◇◇

*Your euphoric, dreamlike experience in the void comes to an abrupt end. After an unknown interval, consciousness resumes in only a fleeting, vague form, similar to waking up from a long, deep sleep.*

*Then, in your mind's eye, vague images begin to appear in the haze. Slowly they become clearer.*

*Initially, you remember the spectacular experience of infinite space and the sourceless light within. You try to recall the events leading up to that fantastic encounter, but your memory doesn't go beyond these most recent events. The amnesia is deeply penetrating.*

*As you regain consciousness, disconnected hints and whispers of your life passively surface to memory in the form of images and emotions. It feels like some kind of reformation process. As it goes on, you slowly regain your identity—your sense of self. It's like you're beginning again, emerging from an infantile state of intellect.*

*Soon your mind floods with the emotions and nuances of a lifetime's worth of past experiences. You remember your life up until your chilling descent into cold delirium. The wave of recollection peaks. Finally, you relive the moments before any of that whole, surreal exchange began, when you wondered whether you'd ever see the world again.*

# 7

# TILL BRAIN DEATH DO US PART

## COLD'S CONNECTION WITH CONSCIOUSNESS

Atlanta, Georgia, 2011. A fifty-five-year-old man was admitted to the emergency-care unit at Emory University Hospital. He appeared lethargic and was suffering from respiratory failure, an observation precisely in line with the paramedics' statement that "he cannot breathe." His dire situation became worse when he suffered a cardiac arrest. After nearly twenty minutes of the ER team's frantic attempts at resuscitation, he started to breathe again and was transferred to an intensive care unit. There, although he maintained a pulse, his condition slowly declined. He was hypothermic and became unresponsive. His shallow breaths grew less and less frequent. He had no reaction to his name being yelled or any other vocal trigger. This patient was in big trouble.

Therapeutic hypothermia was initiated in the hope of slowing his metabolism and protecting his heart and brain tissue from a lack of oxygen caused by an absence of blood flow. His core temperature was reduced to 91°F (33°C). Still he remained unresponsive to vocal prompts. He showed zero reaction to being pinched and prodded. He failed to display any pupil or corneal reflexes to bright light and exhibited no cough or gag reflexes. Then, just when you'd think the situation couldn't worsen, his infrequent breathing stopped completely.

Despite the medical team's efforts, his pulse did not return.

After the steps to maintain cooling were terminated, a neurological examination, undertaken once his body had been rewarmed, determined that the patient was brain dead. To be absolutely sure he had undergone clinical death, after six hours he was given another test for any sign of brain activity. None was found. He was disconnected from the ventilator; life support was removed. His family was contacted with the sad news of his passing. They agreed to the removal of his organs for donation.

The next day, the patient's body was readied for organ procurement. It was wheeled into the operating room and put on a table. Then, to the utter dismay of the staff, in front of their eyes it started coughing. They called the attending neurologist. The cold body was checked for reflexes. The patient's eyes responded to light, and he showed signs of infrequent breathing. He was returned to the intensive care unit, where he was placed on full life support. The ICU staff made plans to observe him over the next several days.

His relatives were contacted and informed of what had happened. When they arrived at the hospital they demanded to know why he could have been declared dead. Was he now really alive? Would he stay alive? How could this happen? Had it happened before?

The main concern was the outlook for his recovery. What kind, if any, could he achieve? Unfortunately, the prognosis wasn't good. Although the patient continued to show some reflexes, and to cough and breathe occasionally, his condition did not improve beyond that. Neurologists who examined him concluded that he was able to maintain only a bare minimum of the most basic brain activity. Reflexive actions like pupil dilation and coughing are controlled by the brain's most evolutionarily ancient, primordial regions, located in the brain stem. Advanced cerebral activity, which in humans has evolved more recently, was absent in this patient, and the chance of his regaining any normal form of consciousness, let alone having any memory of his past or his self-identity, was infinitesimal.

Such minimal, vestigial brain activity might have been preserved because the patient had been cooled. Therapeutic hypothermia, which reduced his brain's metabolism, could have temporarily decreased his neurological activity to a point so low that it dropped beyond any ability to be measured during attempts to resuscitate him. When he was rewarmed, any tremors of activation slowly increased to be sufficient enough to trigger minimal signs of life. Fundamentally, however, the activity in his brain was incapable of increasing further, in either magnitude or complexity. By the point of his apparent resurrection, the greater part of his brain had been too starved of oxygen and nutrients for any hope of further recovery.

Without a high level of function in the major parts of the brain, the patient would have no sensation, perception, or awareness. He would never again exist with any modicum of consciousness. He would have no memory of any event that had ever happened to him. Anything that he had learned—how to walk, talk, eat, and so on—would essentially have been erased. He would remain a vegetative zombie, without the memories, personality, or characteristics that made him *him*. According to all conventions, the implications were dire.

This unfortunate case demonstrates fundamental connections between the brain, the state of cold, and the human mind. Although cold kept the most rudimentary parts of the patient's brain active and functioning, it was too late to save the regions essential for his mind to regain consciousness and memory. The events and their impact on the patient's family brought the issue of hypothermia's ability to mask brain death into discussion within the medical community; since then, doctors have learned to use caution when declaring brain death after hypothermia.

From a broader perspective, the case only scratches the surface of a history of bizarre events surrounding cold brains, life, death, and consciousness.

## TRAVELING WITHOUT MOVING

Many would argue that consciousness, as defined by one's awareness, perception, thought, and memory, can transcend the physical world—and the loss of brain activity. Consider, for example, the widespread conviction that after we die our existence somehow continues in the form of an intangible soul or spirit.

Sure, it's common to believe in an afterlife, but where things get particularly mysterious surrounds what are called *near-death experiences*. NDEs are exactly that: they have been defined as experiences of consciousness that occur when a patient undergoes a cardiac arrest. Typically, they are said to happen during an absence of detectable brain activity and are remembered after resuscitation. That said, if NDEs do occur after death, it is a complete mystery as to how one can achieve awareness and form memories of their experience while their brain is inactive.

Although subdued brain states are often caused *accidentally* by cardiac arrest or severe trauma, they are also *deliberately* induced during open-heart surgery or brain surgery. To be able to operate on these vital organs, an arrest of blood flow is usually essential. Otherwise, any such procedure would be catastrophic; the patient would simply bleed out and die. That's where cold comes in.

When a cardiac arrest patient is cooled, their heart and brain are protected from the effects of a lack of oxygen and nutrients normally supplied by the blood. The reduction in metabolic activity essentially slows time, preventing the onset of damage and allowing surgeons to carry out all the necessary procedures. Upon the completion of the surgery, the patient is rewarmed, and optimally they spontaneously regain a heartbeat. If they don't, they receive shocks that can kick-start the heart into beating again.

The point here is that cold essentially creates the opportunity for an NDE.

Although they are commonly viewed with a healthy dose of skepticism, the similarities between the NDEs reported by people who have been temporarily clinically dead are eerie. Countless records of people describing their NDEs reveal similar types of events. What stands out as particularly spooky is that this remarkable pattern has existed since the earliest accounts of NDEs and is robust despite differences in age and culture. Yet why the similarities exist remains a mystery. Perhaps a scientific approach to explaining them can provide insight into both how the brain works and fundamental aspects of the human psyche.

In 1983, in an attempt to standardize future research, the professor and psychiatrist Bruce Greyson categorized typical NDEs to create a scale that could measure the depth of any such occurrence. It consists of sixteen features typical of the phenomenon, including:

* feeling a sense of harmony or unity with the universe
* perceiving a brilliant light
* entering an unearthly world
* the presence of mystical beings

And perhaps most famously, it includes an "out-of-body experience," in which the patient reports existing without their body. While navigating the "astral plane," some experiencers even claim to have observed real-world events and objects.

For each of the sixteen features, there are three possible answers, each with a corresponding number of points, based on the intensity of the feature. For example: "Did you feel separated from your body?"

* No. (Zero points)
* I lost awareness of my body. (One point)
* I clearly left my body and existed outside it. (Two points)

Out of seventy-four NDE accounts, the average total in Greyson's initial study was fifteen.

One patient, however, stands out for having had a particularly extraordinary NDE. It happened in 1991, while she was in a state of deep hypothermia.

At age thirty-nine, the late Pam Reynolds, a musician from Atlanta, Georgia, suffered a sudden onset of dizziness, trouble speaking, and difficulty moving parts of one side of her body. She was having an aneurysm. After a CAT scan revealed that the massive aneurysm lay deep within her brain, her neurologist recommended immediate surgery. Owing to the aneurysm's size and location, the surgeons had to stop all blood flow in her brain before being able to remove it. The amount of time they needed to perform the procedure, however—during which no oxygen or nutrients would reach the brain cells— was well beyond the six-minute limit before permanent brain damage occurs. The only way to execute brain surgery was to buy time by cooling Reynolds into a state of deep hypothermia, around 59°F (15°C), until her heart stopped beating and the circulation to her brain terminated. Only then would a window of time open for the neurosurgeon to safely remove the massive aneurysm.

The operation took place at Barrow Neurological Institute in Arizona. Carefully, the medical team cooled Reynolds to 60°F (15.5°C), a fatal temperature under uncontrolled conditions in cases of accidental hypothermia. Indeed, she remained clinically dead for around forty-five minutes while the aneurysm was removed. During that time she experienced what is known as *electrocerebral silence*—a flat line on the electroencephalogram (EEG) indicative of an absence of cerebral brain activity. By all forms of measurement, she was deceased.

While in this suspended state, Reynolds underwent possibly the most intense NDE ever recorded; it measured an astonishing twenty-seven on the Greyson scale.

She reported encountering nearly all the features, many at the fullest intensity values. The centerpiece of the experience was her memory of exiting her body through the top of her head and encountering the spirits of a number of deceased relatives in a wash of brilliant white light. She reported a feeling of extreme well-being as well as heightened senses. As she remembered, "At some point very early in the tunnel vortex I became aware of my grandmother calling me. But I didn't hear her call me with my ears. . . . It was a clearer hearing than with my ears. I trust that sense more than I trust my own ears." Everyone whom she saw in the vision somehow appeared to know that her time there was only temporary, and, according to her recollection, somewhat against her will she was returned to her physical body.

What really stands out about Reynolds's NDE is the degree of awareness she claims to have maintained while unconscious; according to her recollection, it somehow penetrated into the real world. Despite being dead, she recalled perceiving what was happening around her in the operating room during her own brain surgery. The accuracy and detail of her descriptions triggered shock and disbelief in the members of the surgical team, leaving them deeply unsettled.

She recounted four events—mostly from auditory memories— that were later verified with the medical staff present during the procedure. The first was the sound of the drill boring a hole into her skull: "It was a natural 'D'"—that is, the musical pitch. "As I listened to the sound, I felt it was pulling me out of the top of my head."

Next, she accurately described the unpleasant noise made by the bone saw: "I heard the saw crank up. . . . It was humming at a relatively high pitch and then all of a sudden it went *Brrnrrrrrr*, like that."

She was able to recount specific parts of the conversation taking place around her. The cardiothoracic surgeon pointed out that on Reynolds's right side, the arteries intended for integration with cardiac bypass tubing were too small: "Someone said something about my veins and arteries being very small. I believe it was a female voice and that it was Dr. Murray, but I'm not sure. . . . I remember thinking that I should have told her about that."

The fourth recollection is perhaps the most surreal: "They were playing 'Hotel California' and the line was 'You can check out anytime you like, but you can never leave.' I mentioned [later] to Dr. Brown that that was incredibly insensitive and he told me that I needed to sleep more. [Laughter.] When I regained consciousness, I was still on the respirator."

Reynolds described characteristics of the bone saw that was used, apparently without having seen one previously. "The saw-thing that I hated the sound of looked like an electric toothbrush and it had a dent in it, a groove at the top where the saw appeared to go into the handle, but it didn't. . . . And the saw had interchangeable blades, too, but these blades were in what looked like a socket wrench case." These details were verified by Robert Spetzler, the head neurosurgeon who operated on Reynolds. He had pioneered key aspects of the hypothermic circulatory-arrest technique that was used during her procedure.

How was it possible that she could know these things? Did she indeed transcend her physical self while in a frigid state? Did she really "pop out of the top of her head" in some incorporeal form, as she claims to have done? Although Spetzler remains perplexed by the matter, other medics have offered up theories.

According to Gerald Woerlee, an anesthesiologist who has extensively researched NDEs, Reynolds experienced brief moments of hazy, disjointed consciousness during the surgical procedure. He explains that although she was under the influence of a combination of narcotics that induced general anesthesia and paralysis, she was not actually in a state of clinical death during the specific events she recalled. Indeed, there are a number of case reports of patients who, after being given anesthetic, experienced consciousness and perceived the events around them during major surgical operations. That may sound horrific, but because the

patients were high, confused, numb, and apathetic, the phenomenon was probably similar to having a strange dream. (Incidentally,
I recall waking during a surgical procedure to remove all four of
my wisdom teeth, despite the local and sedative anesthetics that
had been given to me. Afterward, I had a dreamlike recollection
of opening my eyes and asking the dentist if he could "move the
mirror so I can see better"—a request that my dentist confirmed,
amusingly, as having actually occurred.)

Specifically, Woerlee suggests that both the noise from the
tools and the doctor's comments about Reynolds's veins occurred
prior to her entering electrocerebral silence. She heard "Hotel California" playing after the removal of the aneurysm and after she
had regained heart and brain activity, while her temperature was
approaching normal. An insufficient amount of sleep-inducing
effect from the pharmaceutical sedatives intended to render her
unconscious could account for Reynolds's sense that she had left
her physical body.

Still, her detailed visual description of the bone saw raises an
eyebrow because it was impossible to see the tool from her position on the operating table, let alone while her eyes were covered
with tape, a routine aspect of the procedure. In fact, Reynolds
is not the only person to accurately recall real-world events and
objects during a period of electrocerebral silence. Other patients
have even described events and objects in remote locations.

A famous example involves "Maria" (not her real name), who
in 1977 suffered a cardiac arrest at Harborview Hospital in Seattle. She had an NDE during which she traveled to the roof of the
hospital while her physical self remained in the operating room.
There, she was surprised to find a tennis shoe. Some days after her
operation, she told her critical-care social worker, Kimberly Clark
Sharp, about the strange out-of-body experience and described
the shoe in detail: "Its little toe area was worn and one of its laces
was stuck underneath its heel." Skeptical but curious, Sharp could
not resist checking for the shoe. To her astonishment, she found

it in the exact location identified by Maria and in the exact condition she had described, with a worn toe and a stuck lace.

This anecdote, however, is just that: a story with no proof. The same goes for Pam Reynolds's experience, or any other claim of witnessing real-world objects and events in the absence of an active, functioning brain. As far as what these events owe to therapeutic hypothermia, cold has opened a new pseudoscience characterized by a lack of evidence. So how would it be possible to scientifically confirm or deny the ability to perceive events while flatlining?

◇◇◇◇◇◇

THE RESULTS OF an in-depth study, ongoing since 2014, are some of the most anticipated in history. Its moniker spares no exactitude. It's called AWARE II: AWAreness During REsuscitation— A Multi-Centre Observational Study of the Relationship Between the Quality of Brain Resuscitation and Consciousness, Neurological, Functional and Cognitive Outcomes Following Cardiac Arrest. It is headed by Sam Parnia, a British professor of medicine at the New York University Langone Medical Center, where he is also director of research into cardiopulmonary resuscitation. AWARE II expands on the original AWARE study. Ironically, although that study was unable to determine if perception could occur during an absence of neurological activity, it did reveal that as many as 46 percent of 2,060 cardiac arrest events involved memories of events perceived as occurring during the cardiac arrest.

Previous investigations have shown that patients who have NDEs seem to enjoy better outcomes, physically and psychologically. The running hypothesis is that for various complex reasons, still to be discovered, perhaps those patients are less affected by an absence of blood oxygen and nutrients during cardiac arrest. Understanding the relationship between NDEs and brain activity patterns could ultimately lead to better recoveries and to improved long-term outcomes.

But what about the out-of-body experiences and claims of real-world perception? The original AWARE study attempted to answer this central question. Cards with either the names of widely known public figures or newspaper headlines written on them were distributed in emergency rooms. They were placed facing up and visible only from vantage points near the ceiling—on top of a tall cabinet, for example. Only 22 percent of the NDE descriptions examined in the study were from cardiac arrest events in operating rooms or emergency rooms where a shelf-card had been placed. None of the reports included a correct identification of what was written on the card in the small number cases where there was one in the same room as the event.

The AWARE II study uses electronic tablets to present imagery and sounds while patients experiencing cardiac arrest undergo resuscitation. Afterward, patients who agree to an interview are asked to describe any memory they retain from when they were temporarily dead. If, like Pam Reynolds and Maria, they recall visual and auditory events that are in line with what actually occurred around them, correctly identifying the stimuli provided by the tablet, it would be a step toward confirming whether conscious perception can occur during an absence of brain activity. As of the writing of this book, data collection is nearing completion.

The idea of consciousness transcending known physical laws in the manner described by those who recall such experiences during NDEs can be described as, well, extremely unlikely. The mind is typically seen as having an intrinsic connection with the brain, a physical object. Yet the nature of that connection has always been one of humanity's greatest mysteries. Considering that scientific evidence is always materially or physically based, it may never be possible to prove beyond a doubt that an out-of-body experience has actually occurred. Perhaps this is why these stories perpetuate. NDE-ers will simply have to learn how to manipulate tangible objects! At least, with significant help from cold, AWARE II opens the possibility of gathering evidence that will be difficult to refute.

## BRAIN FREEZE

You are unique, one of a kind. Congrats!

As far as we know, your internal experience of that uniqueness—what it's like to be you instead of someone else—is entirely supported by the uniqueness of your brain. That means, unlike many of your other organs, it is irreplaceable. You simply can't swap it out if it becomes damaged. No chance. Sorry.

If the brain in your head were replaced with someone else's, your memory, personality, identity, and sense of self—that is, you—would no longer exist. It's more accurate to say that instead of you having a brain transplant, whoever donated the brain would have a new body. Similarly, if your head were to be replaced by another, the new Franken-person would not be you. Rather, the head, as a seat of consciousness, would have a new body.

Indeed, one of the most fascinating yet bizarre sciences fundamentally supported by cold is head transplantation. Far from science fiction, the possibility, with humans, currently exists, according to a community of surgeons and neurologists. The procedure would require surgically decapitating both a donor and recipient and transplanting the recipient head onto the donor body. Ideally the donor would have suffered brain death but otherwise be healthy for a successful body procurement.

The rationalization is that this could be a life-saving procedure for those with terminal illnesses that affect the head and body in different ways. If it were possible to receive a new, functional body, those suffering from amyotrophic lateral sclerosis (ALS) or spinal muscle atrophy, for example, could live and possibly regain use of their limbs. The idea is fraught with controversy over procedural, psychological, and ethical issues that currently remain largely unresolved. Cold is at the center of the controversy, because it is basically an enabler; the head and body to be joined must be cooled to prevent damage from occurring during the lengthy period of cardiac arrest that would be required during the operation.

The concept of a human-head transplant is not novel, and, yes, the earliest reference I can find comes from *fictional* literature—*Professor Dowell's Head*, a 1925 sci-fi novel penned by Russian author Alexander Belyayev. By the 1950s, the imaginary procedure had become a reality—using dogs. In a series of bizarre, unethical experiments, Vladimir Demikhov, a noted Russian surgeon who pioneered organ transplantation, connected the heads and forelegs of donor dogs to host dogs' circulatory systems. The creatures were able to move and react to various stimuli. Thankfully for future animals, though, by the 1960s the head-transplant experiments were plagued by worldwide criticism for their inhumanity and cruelty.

Nonetheless, in terms of the technical feats, the experiments garnered admiration from prominent surgeons around the globe, including the late American Robert White. Although he's credited with making key insights into using cold as a tool to enable organ transplantation, "Humble Bob," as he nicknamed himself, essentially became the new Demikhov. And the new Dr. Frankenstein.

White, who was born in Minnesota in 1925, graduated cum laude from Harvard with a medical degree in 1953. After accepting a position at the prestigious Mayo Clinic, located in his home state, he began experiments to cool the brain and spinal cord for surgery after trauma, to determine if cooling could protect the organs by lowering their metabolism while they were cut off from blood supply. Researchers were able to cool the spinal cords of monkeys to as low as 43°F (6°C) for four hours and still demonstrate the restoration of nervous system functionality upon rewarming. Owing to the tissue-preserving power of cold, their success started a chain of experiments that would change medical history and open new frontiers.

White was motivated by aims that went beyond the treatment of trauma. He began to use hypothermia in the removal and preservation of the functioning brain—alone, with no body. In 1963, in the confines of his laboratory, he removed the brain of a monkey while keeping it active with blood flow. His experiments are well documented on film. An isolated monkey brain pulsates in a metal frame while surgically connected tubes circulate blood through its intricate vasculature. In a follow-up experiment, he connected the brain by wires to an EEG. Horrifically, the EEG readout showed activity. The monkeys, existing only as bodiless brains, remained essentially alive for up to three hours. The experiments, which were continued over the rest of the decade with both monkeys and dogs, were later extended to forty-eight hours.

Were these brains conscious? Did they retain the capacity for emotions? Did they experience extreme pain?

White displayed little concern for the monkeys' well-being; he was insatiably curious about any level of awareness the animals retained. The next step was to repeat the operation, but with the whole head included. By 1970, he was working as a professor of neurosurgery at Case Western Reserve University School of Medicine. There, with a team of around thirty doctors, nurses, and technicians, he surgically decapitated two monkeys. The recipient monkey's head was connected to the circulatory system of the donor monkey's body.

Again, the EEG began to show neural activity in the head-transplanted monkey. Soon it opened its eyes and looked around. The Franken-monkey tracked the movement of objects, responded to being touched, and was able to bite, chew, and swallow. But lacking any spinal cord connection, it was unable to move below the neck. The monkey lived for thirty-six hours. In White's words from a 2007 interview, "He was not a happy monkey."

This was only the first in a series of experiments with primates conducted by White. The longest-surviving monkey was kept alive for about eight days, owing largely to careful application of hypothermia.

As you can rightfully imagine, White was bombarded with criticism. His colleagues considered the experiments barbaric. In the press, he was referred to as "Frankenstein" and "Dr. Butcher." He received threats by mail and by phone, as did his wife and some of his ten children. Still, White continued his work, pushing scientific and medical frontiers for the benefit of patients with debilitating spinal cord illness.

In fact, White, a devout Christian, served as the leading ethical advisor to the Roman Catholic Church. He personally consulted with Pope John Paul II on a number of occasions, assisting the church on controversial topics such as organ transplantation, in vitro fertilization, and brain death.

Some Russian and Chinese doctors deeply admired White's accomplishments. He visited both countries to train physicians and neurosurgeons. In Russia, where dog experiments had continued, he witnessed two-headed dogs. The Soviet authorities invited him to stay and continue his head-transplant experiments, but White lost any potential interest upon discovering that their primary motive was achieving human immortality. Although he regarded the brain as the repository of the human soul, and he deemed cold essential for keeping it operational while it was disconnected from a body that could supply it with blood, he didn't share their idealism.

White faced reproach for his treatment of animals until his death. Nonetheless, he had a penchant for exchanges with the media, participating in a steady train of interviews until 2009, the year before he died. He was a true polemicist, having no fear of engaging with his critics despite threats of physical violence. He was equally known for his advancement of lifesaving surgical practices. His work was way ahead of its time, and his impacts on neurosurgery and spinal cord surgery stand out today. According to Paul Bucy, editor of the journal *Surgical Neurology*, they were seen as "among the most outstanding contributions in the last quarter century. His work was revolutionary in the broadest sense." White was nominated for a Nobel Prize twice, in 2004 and in 2006.

During White's lifetime, transplantation of a human head was never undertaken. Famously, he claimed that the procedure could cure Stephen Hawking, but despite his enthusiasm, difficulties surrounding immune system rejection and spinal cord reconnection remained insurmountable. Nonetheless, he believed that it would inevitably happen, saying, "What has always been the stuff of science fiction—the Frankenstein legend, in which an entire human being is constructed by sewing various body parts together—will become clinical reality in the 21st century."

∞∞∞∞∞

HUMBLE BOB'S PREDICTION has never been closer to becoming a reality. The twenty-first century has yielded a refreshed, emboldened desire to perform the first human-head transplant. The motivation for this bizarre achievement is held by a growing group of surgeons worldwide, and at the forefront is Sergio Canavero.

Born in 1964, in Turin, Italy, where he attended medical school, Canavero has been practicing neurosurgery for over twenty years. He is widely published and has made influential contributions to pain research and to the treatment of Parkinson's disease. Besides being a distinguished neuroscientist, he holds some strict personal philosophies regarding abstinence: no television, no driving, no drinking alcohol, no beef. Women? Yes. He has actually published a book on the art of seduction: *Women Discovered*. He's also well practiced in martial arts.

In 2013, Canavero founded an ongoing project called HEAVEN, a loose acronym for *head anastomosis venture*, with the aim of achieving the first successful human-head transplant. ("Anastomosis" is defined as a cross-connection between anatomical systems.)

But why *heaven*? Surely undergoing the procedure would be more like hell.

Similar to the Russians who tried to entice White into joining them, Canavero believes that head transplants will ultimately unlock the door to immortality. He envisions a future when human clones are harvested for younger, healthier bodies onto which people's heads can be transplanted when their physical condition declines naturally, or upon serious illness. The newly cloned body would be similar in stature to that of a twenty-year-old, but would be grown in the span of a year. Canavero clarifies that the bodies would never be made conscious, so that when it came time to decapitate them, "it probably would not be murder."

Canavero asserts that by using therapeutic hypothermia, along with advances in nerve reconnection, he could successfully perform a human-head transplant today. Hypothermia would be essential, crucially preserving both the recipient head and donor body during the reconnection process. The estimated cost is up to $20 million.

Canavero had his first volunteer, a Russian computer scientist named Valery Spiridonov. Before the actual procedure could be undertaken, however, Spiridonov changed his mind. Born in 1986, Spiridonov suffers from Werdnig-Hoffman's disease, a terminal condition that causes progressive muscle atrophy. It has rendered him incurably quadriplegic. Spiridonov backed out because he fell in love with the woman who is now his wife and they had a child together. She was against the procedure, and her opinion weighed on him. Meanwhile, his condition seemed to have stabilized, and he did not want to spend months in rehabilitation, away from his family.

There are good reasons for deciding not to undergo such an outrageous procedure that are, however, more generalizable. Primarily due to ethical concerns, the surgery has not yet been tested on primates. No exact model has been developed for how it could or should happen because researchers and surgeons have no idea whether it could actually work—a less than ideal context for testing. Moreover, surgeons, doctors, and bioethicists widely share the opinion that the science behind any proposed version of an actual protocol is shoddy and unsupportable. Needless to say, the ethical considerations are complex and unresolved. The upshot is that if a head transplant were performed today, critics feel that the patient would likely end up dead, if not paralyzed in another body, severely demented, and psychologically traumatized.

A primary obstacle is that such an operation would require the largest surgical team ever assembled for a single operation, as well as unprecedented levels of cooperation and coordination among the team members. They would need to simultaneously decapitate both the recipient and the donor and then maintain the respective body and head at ideal hypothermic temperatures. They would surgically connect the donor's body to the recipient's head as fast as possible before serious damage could result from a cessation of blood flow. The optimal temperatures remain unknown, but it has been proposed that the head should be kept under profound hypothermia, at around 50°F (10°C), while the body is kept at a relatively higher temperature. During the reattachment, the most glaring unknowns concern the integration of the donor's and recipient's nervous systems via the spinal cord. The severed axons will remain stable for only ten to twenty minutes before starting to die. Maintaining or growing connections is absolutely crucial if the body is to recover any functionality. Upon recovery, the recipient can only hope to gain control of vital process and to be able to move the new body beyond the spastic muscle contractions likely to be initiated.

And yet, even if a successful transplant enabling movement in the donor body is achieved, the potential for psychological catastrophe remains a central concern. The contention surrounds a widely researched theory regarding mind-body interaction called *embodied cognition*. It holds that central aspects of our mind are dependent on physical characteristics of our entire body and on how we experience the world through action and perception;

one's mind shares an intrinsic connection with not only the brain but the whole body. Our experience, as achieved through our body, is central to the memories we form throughout our lives. These memories compose our sense of self, and thus our physical experience is essential for maintaining personal identity.

Imagine relaxing at the beach under the sun, forgetting the world around you, nodding off, and then waking up to find that your body is not *your* body. You realize that you have no choice other than to use this body for everything that you need and want to do for the rest of your life. There's no escaping it. Potentially, there could be no accepting it either.

If we suddenly were given a new body that was much taller, shorter, rounder, or skinnier than our current body, we certainly wouldn't feel like ourselves. We could require a drastic shift in self-identity. Or if we became considerably more muscular, our abilities would also change accordingly. What if the donor body had some special physical ability, for example as a professional athlete or musician? What would it be like to suddenly have that body? No matter how physically similar the donor is, their body could still *feel* entirely different, simply because it is.

Besides these procedural and psychological issues, a host of social concerns also arise. Who could receive a head transplant? Under what circumstances? Most agree that those with terminal illness should be given priority, but if it becomes a successful practice, would only those wealthy enough to pay for the operation get a new body? Would it mean the return of resurrectionists—grave robbers who harvested bodies for anatomists in the eighteenth and nineteenth centuries? Would those who could afford to do so be entitled to keep transplanting their head onto a new body whenever their demise approached, enabling them to live indefinitely?

Another central issue involves offspring. If your head were transplanted onto the body of someone unrelated to you, which would likely be the case, any successful act of procreation would result in a child that you are not related to. Instead of being able to have a child of your own, you could have only the child of your partner and the donor. Ethical and practical concerns regarding the child's relationship with the donor's family would arise.

I've listed here only the issues that have actually been realized so far. No doubt many ethical, psychological, and social concerns surrounding human-head transplantation remain to be discovered.

The prospect of the surgery as supported by hypothermic cooling has been catapulted to the fore nearly entirely by Canavero and his unflappable confidence in the procedure's possibilities. Since Spiridonov offered himself as the first subject in 2015, academic publications as well as popular media articles on the topic have increased drastically. Heart transplants, hand transplants, and face transplants once teetered on the ethical fringes, igniting deep-rooted controversies and pushing borders. Perhaps head transplants, like those procedures, will someday be accepted by those who are currently skeptical.

## "VITRIFIXATION"

If you can imagine, an even more cutting-edge means of using cold for radical life extension is now in development and making headlines worldwide. The brainchild of MIT scientist Robert McIntyre, "vitrifixation" (a neologism of McIntyre's), ironically, is guaranteed to be fatal. Don't ask McIntyre about immortality, though. Rather cleverly, he asserts that vitrifixation is fundamentally aimed at indefinitely preserving "living memories." To understand what he's talking about, one needs to read between the lines.

A California native born in 1989, McIntyre was inspired by audio recordings he made as a youngster of his grandmother. He recorded her talking about moving from Oklahoma to Texas by covered wagon, among other antiquated experiences, all of which fascinated him. Hopeful but naive, McIntyre considered the possibility of preserving a memory so that it could one day be reexperienced by someone else. After exploring the neuroscience involved, he became aware of how outlandish this may seem relative to our current knowledge and capability. He decided to study artificial intelligence with the belief that if this feat is achievable, it would require extreme computational endeavors.

Meanwhile, accumulating evidence was beginning to show an essential role of the entire brain's neurally connective structure for encoding, storing, and recalling memories, known as its *connectome*. Now it's widely accepted that memory is not stored in any single part of the brain, but rather in the strengths of connections between neurons and the relative timing of neural activations. Memory is one of our brain's greatest accomplishments.

Think about it. Your memory is central to your identity and personality. What it's like to be you, as opposed to someone else, is fundamentally created out of your unique experience in the world. As you learn and grow, those experiences change the structure of your brain, making an indelible imprint. Essentially this is memory, and in many ways, essentially *you* are memory. If you can preserve the brain's current connections, you can preserve the imprint of a lifetime of experience, and thus you can preserve you. At least in theory.

Following a life-changing talk with his father about biology and preservation, McIntyre began to volunteer at 21st Century Medicine, a privately funded cryobiology lab in California. There, after teaching himself the fundamentals of neuroscience, cryobiology, perfusion, and microscopy, he began collaborating with Greg Fahy, vice president and chief science officer of the company. In an attempt to accurately preserve brain tissue, McIntyre tested nearly one hundred chemical formulations before making a life-changing discovery.

He found that a solution called *glutaraldehyde* could do the job with astonishing fidelity. Glutaraldehyde is nothing new. It's been used since the 1960s as a disinfectant, as a preservative, and for the treatment of warts. What is new is the discovery of its incredible capacity for holding neural synapses—the connections between neurons—in place while keeping them from degrading. When McIntyre and his colleagues carefully perfused it through the entire vascular system of a pig's brain, they were surprised to find that the fixed tissue appeared to remain synaptically intact, based on an examination with electron microscopy. In a case of hi-fi memory preservation, the entire connectomic structure, consisting of trillions of connections within the pig's neuronal network, remained in place.

Glutaraldehyde causes the brain's molecular structure to adhere together in a manner similar to the preservation achieved via formaldehyde; as a fixative, it essentially solidifies the brain's connective structure. Then the brain is vitrified using ethylene glycol—antifreeze, but in a more concentrated solution than what you use in your car radiator—which prevents the formation of ice crystals. Because vitrified materials are not solid, they are susceptible to slow deformity and decay over time—unless they are frozen. That, of course, is where cold comes in. Freezing a vitrifixed brain at temperatures of −187°F (−122°C) or below slows atomic motion enough that the brain's connectome can be preserved for hundreds of years.

These advancements won McIntyre and Fahy the $100,000 prize awarded by the Brain Preservation Foundation, the nonprofit mentioned in Chapter 6 that is specifically devoted to long-term whole-brain storage with intact synaptic connections.

Currently, McIntyre is independently researching vitrifixation via his own startup, a company called Nectome. The name is a portmanteau of *necto*, loosely Latin for "I bind myself," and *connectome*. Nectome's potential to achieve high-fidelity brain preservation has stirred considerable interest from funders. So far the firm has received nearly a million dollars in research grants from the National Institutes of Health, "to enable whole-brain nanoscale preservation and imaging, a vital step towards a deep understanding of the mind and of the brain's diseases." It has also received an undisclosed amount from the venture capital firm Y Combinator, the same company that provided startup funding for Airbnb and other successful businesses.

McIntyre has been careful to keep his work grounded in scientific reality. Yet he has simultaneously managed to leverage the idea of memory preservation and immortality to create levels of excitement and contention unparalleled since the early days of cryonics.

Although the logic that the brain's connectome may be essential and therefore preserving it must also preserve memory is basically sound, the controversy surrounds the extraction and revival of consciousness and the prospect of immortality. Unlike cryonics, which assumes the future reanimation of one's physical body or head via nanotechnology, vitrifixation is entirely toxic. Once a brain is vitrifixed, although its structure remains intact, it is most definitely poisoned beyond any hope of physical revival. Thus, instantiating the memories—the preserved "self"—requires the ability to deduce the molecular structure of a preserved brain and then to transfer the information to another, nonbiological medium. There, the idea is that its activity can be simulated or replicated and the memories reexperienced. If indeed you are essentially your memory, and your memory is intrinsically tied to the activity of your connectome, the hope is that you'll make a return, despite the absence of every single atom that is currently a part of you as a physical entity.

Many advocates of the technology are transhumanist futurists who believe that a mind can consist of software run on advanced artificial hardware. By "advanced," I mean software that can simulate the goings-on of the brain's 150 trillion synaptic connections; indeed, we're a long way from such computational fortitude. Others assume that a new brain-like physical system will be created that can reinstantiate the brain's neural processes and govern a robotic body. Their reasoning is that we already replace broken body parts with prosthetics, which now even include neural

devices that function in place of neurons. Currently, neural prosthetics exist as artificial brain stimulators, commonly used to treat
parkinsonism. If it's possible to replace parts of the brain's neural
structure, it will only be a matter of time until you can replace *all*
of it.

Although the prospect sounds outrageous, it has backers with
some pretty serious cred. The late Stephen Hawking believed that
within a few centuries at most, this could be accomplished. He
said, "It's theoretically possible to copy the brain onto a computer
and so provide a form of life after death"—although he admitted
that we are presently nowhere near such a capability. Other scientists, however, are less optimistic. Richard Brown, a neuroscientist at Dalhousie University, in Nova Scotia, who is working on
memory, asserts, "When you die, the brain cells die, so there is no
memory after death. . . . It is not possible to find memory in dead
neurons."

Despite the disagreement surrounding the question of exactly
how to reinstantiate preserved memories from vitrifixed brains,
Nectome has already preserved a human brain. Ideally, it holds
the memories of a woman from Portland, Oregon, who died of
natural causes. The six-hour process began two and a half hours
after her death; according to McIntyre, her brain is "one of the
best-preserved ever." Still, that window of time was enough for the
brain to incur considerable damage; therefore, it is being used to
examine the results of the preservation procedure rather than being cooled for century-scale intervals and future reinstantiation.

News of these events created enough of a buzz that vitrifixation is increasingly being considered as an alternative to cryonics. Advocates believe that despite the physical impossibility
of reviving one's actual brain, there is no physical or theoretical
barrier preventing the reinstantiation of the brain's connections
on an artificial substrate. They argue that a high-fidelity preservation of the connectome opens the possibility of retaining enough
connection-related information to make a reanimation possible.

The vitrifixation procedure could one day be a well-accepted option for those diagnosed with terminal illness.

For this to happen, however, considerable legal hurdles must be surmounted. First, for the process to work, the patient must be alive when it starts and, ideally, already chilled as much as possible. Only in this way can the state of being cold slow metabolism and minimize brain damage from a lack of oxygen. The legal implications of homicide are obvious. Yet there may be a workaround in the form of doctor-assisted suicide, which is now legally allowed in some states, including California. If these legal obstacles are overcome, the next issue involves scientific justification: Does the procedure actually work on a *human* brain?

Ken Hayworth, president of the Brain Preservation Foundation, which awarded the cash prize to McIntyre and Fahy, is a full believer in the potential of vitrifixation to radically extend life. Yet even he said in a 2018 interview, "If you take that shortcut and start offering it prematurely then you may be responsible for thousands of lives lost. Do not do it. Do not try to do an end-run around the standard medical procedure development process or the medical ethics debates."

McIntyre agrees, explaining that vitrifixation has a long way to go before any clinical testing can occur and that its development should never attempt to circumvent the skepticism and validation of scientists and ethicists.

◇◇◇◇◇◇

*As you open your eyes, your first breath is a deep gasp for air. At first nothing in your vision is clear, but slowly a white ceiling comes into focus. You turn your head to see more. Oddly, as you move you notice a quiet, mechanical whirring, just loud enough to break the silence. You see that you are in a small, featureless room. It's immaculately clean. It seems that you're in a hospital, yet, strangely, there is no recognizable medical device anywhere. Perhaps this is some kind of institution?*

*You're covered in a thin white sheet. You lift your arms, and again you hear another barely audible whirring sound. You study your hands in sheer amazement that you are still alive. You notice that they seem different: they look twenty years younger. You're in disbelief.*

*You take in more of this strange room. You perceive some force restricting your head movement. You are in no way prepared for what happens next. You reach behind you and feel something foreign. It feels like a cluster of cables. You follow them with your hands as they lead toward the top of your head and are astonished to find that they connect directly to your scalp. It feels as though they actually run into your head. On the other end they connect to a metallic sphere at your bedside.*

*You try to exclaim, "Oh my God!" You have difficulty speaking. You fail to coordinate the movements needed to execute the words. Your garbled voice sounds artificial and computerized. To hear more, you vocalize your next thought. "What is going on here? Where am I? What am I?" As you speak, your voice sounds slightly clearer and more normal—like it's somehow being re-created online.*

*Although you don't know what time, day, month, or even year it is, with an epiphany you realize you must be in the distant future.*

*You made it.*

*And you're warm.*

# 8

# REAWAKENING AT 98°F (37°C)

## *A WARM REALITY IN COLD SCIENCE*

The fascinating prospect of using cold to transcend the limits placed on our existence by time and space, from which a glimpse of immortality has imprinted stars in the eyes of exuberant optimists, keeps the delicate field of cold research on the fringe of science fiction and the occult. Yet therapeutic hypothermia has also proven itself an invaluable clinical tool, based on hard, scientific evidence firmly grounded in reality. Fortunately, the medical achievements yielded from a rich history of brave, insightful explorers and their progressive experiments over centuries of investigation have triumphed over an equally long history of pseudoscience and deeply rooted stigmas involving both human and animal cruelty and abuse. Today, the therapeutic potential of cold is understood at an unprecedented level.

That said, there is a long way to go before clinical guidelines and routines for therapeutic hypothermia are firmly established. In many ways, we still have more questions than answers. Recent clinical investigations, however, are moving the practice forward at a greater pace than ever. Today's explorers of cold are as admirable as their predecessors.

## DR. FREEZE

One noteworthy practitioner of cold is Joseph Varon, aka "Dr. Freeze." Originally from Mexico, he is a cardiovascular and pulmonary disease specialist in the temperate climes of Houston, Texas. He has induced hypothermia in over three thousand patients, employing everything from bags of ice to the latest in vitro methods. Accolades about his achievements culminated in his becoming a member of the Royal Society of Medicine in 2013. He's also a renowned worldwide lecturer on how to utilize cold in the operating room.

Among those thousands of chilly patients, some cases stand out for the reknown they've brought to cold as a therapeutic tool with the potential for saving lives. One famous case involved a Canadian vacationer in Mexico back in 2005. Dan O'Reilly was in Ixtapa with his wife and two kids. While at the beach, he went out for a swim and was clobbered by the crest of a large, powerful, and unexpected wave. It sent him crashing head first into the sand, tacoing his spine and severely damaging the connection to his brain that supports respiration. His lungs filled with salt water. His horrified children noticed his distress only when he washed ashore, seemingly lifeless. He was administered CPR for a staggering forty minutes, with no noticeable resuscitation.

O'Reilly was taken to a local hospital, where he was intubated. The doctors decided that the best treatment he could receive was not there, but in Houston, under the care of Varon. He was flown to St. Luke's Hospital, and by the time he arrived he had been hovering at death's door for about three hours. A quick prodding revealed that O'Reilly didn't have a pupil dilation response; even the most basic brainstem activity was absent. His only sign of life was that once every minute or so, he breathed an extremely faint breath. It was just enough for Varon to decide not to pull the plug on the respirator. Varon's immediate assessment was that O'Reilly had about a 1 percent chance of survival.

The first objective was to cool him down to prevent as much brain damage as possible. O'Reilly was wrapped in purpose-built blankets containing a network of tubes circulating chilled fluid, which reduced his body temperature to 90°F (32°C), where it was kept for three days. The big question was whether or not the cooling would slow his metabolism enough to save his brain tissue from further damage, while allowing what little oxygen was left to support activity involved in repairing the already damaged parts.

When he was taken off the cooling regime, his temperature was slowly increased to normal over the span of a couple of days. A nurse noticed that O'Reilly's eyes were open, a tremendous sign of hope to his family and friends—and Varon. Now, however, there was a significant risk that he would be permanently locked in a vegetative state.

The eye opening was the first step in a long, arduous process of recovery. After undergoing an operation to correct the compression in his spine, incredibly, O'Reilly appeared to be cognitively intact. He had his wits about him. The motor-related brain damage he'd incurred, however, required months of physical and occupational therapy before he could move in any normal manner.

O'Reilly's recovery, a victory for both Varon and the use of therapeutic hypothermia, made headlines worldwide. Later that year, when Pope John Paul II suffered a cardiac arrest, his medical team immediately contacted Varon with a plan to fly him to the Vatican to use hypothermia on the pope. Before Varon could board the plane, however, the pope had passed away.

Then, in 2013, when a stroke victim at his hospital risked neurological damage, Varon decided to use therapeutic hypothermia to treat the patient. This time, however, the patient was *him*.

Varon had what he thought was just an ear infection. When the symptoms worsened, he elected to undergo an MRI brain scan. To his astonishment, what he had suffered was a stroke. He decided he needed to get cool.

Employing a combination of cooling blankets and a "cryohelmet," often used on football players who sustain head injury, Varon coordinated his own chilling. He ended up spending weeks in the hospital but recovered after speech therapy and physiotherapy. Cognitively, he survived the event fully intact. He says, "I'm living proof this technology works. I had a . . . stroke and used the helmet immediately. I'm alive." Now, when he treats other stroke victims, Varon is able to encourage them based on his own experience.

## CURRENT CONTENTIONS

Finally, after ages of triumphs and pitfalls, the practice of therapeutic hypothermia is establishing itself in a wide range of clinical applications: to treat asphyxia in newborns, intracranial swelling, severe seizures, spinal cord injuries (like the one sustained by O'Reilly), and kidney and liver failure. The most frequent reason for cooling patients, however, is to provide neuroprotection after a cardiac arrest. Still, the practice remains uncommon and is considered by many as experimental. In many respects, it is. Overall, there's simply not enough research to support standardized cooling procedures; without clear guidelines, doctors would rather use conventional methods than be hypothermia cowboys. In the United States, therapeutic cooling is practiced in only 17 percent of hospitals nationwide.

What has complicated the establishment of a comprehensive methodology is that over the past twenty years, studies published in prominent journals have provided conflicting results about the effectiveness of cooling. Within the publications that do advocate the procedure, disagreement persists regarding a single yet crucial factor: the optimal temperature to which patients should be cooled.

In 2002, two landmark studies, both published in the esteemed *New England Journal of Medicine*, catapulted cooling from a highly controversial practice to one worthy of serious consideration. They advocated lowering the body temperature of cardiac arrest victims to within the range of 90°F (32°C) and 93°F (34°C). The enthusiasm lasted until 2013, when a new study, appearing in the same journal, advocated a temperature of 97°F (36°C), just below the normal human body temperature of about 98.6°F (37°C). This standard would require a considerably different cooling protocol. Faced with conflicting evidence, doctors preferred to use conventional noncooling methods to treat cardiac arrest patients, thereby avoiding the consequences of potentially applying the wrong temperature.

Since 2017, a new guideline put forth by the American Academy of Neurology has prevailed. It sides with the initial two studies that advocated a temperature in the range of 90°F (32°C) to 93°F (34°C). Stated simply, *two* separate studies showing improvements for patients by using this range provide more evidence than one study supporting an optimal temperature of 97°F (36°C).

Clear answers to many other protocol issues remain elusive: How long should patients be cooled? How fast should they be cooled? How fast should they be rewarmed? What is clear are the numerous risks associated with reducing a human's body temperature to the low end of the prevailing optimal range, at 90°F (32°C), including changes in brain and heart rhythms, heart attack, imbalanced electrolytes, hyperglycemia, diuresis, increased bleeding, impaired immune function, and adverse oxygen demand caused by shivering (ironic, since one of the main reasons for cooling is to lower oxygen demand). Thankfully, many of the risks can be prevented or mitigated pharmacologically, and by rewarming the patient very slowly.

## DEEP HYPOTHERMIC CIRCULATORY ARREST

These risks are most strongly linked to *mild* hypothermia—that is, a body temperature above 90°F (32°C), which is commonly used to provide neuroprotection for victims of cardiac arrest. Imposing significantly colder temperatures—down to 64°F (18°C), or three times as cool—is also common practice. The procedure, known as *deep hypothermic circulatory arrest* (DHCA), kills people before it saves them. Indeed, Pam Reynolds (from the previous chapter) owed her famous near-death experience to the circulatory arrest induced while surgeons operated on her brain aneurysm.

DHCA is needed when constant blood flow would otherwise render a surgical procedure impossible. For example, during open-heart surgery or brain surgery, doctors can operate only while no blood is circulating. Cooling the patient's body makes this possible under conditions that render patients unable to breathe by slowing metabolism and thereby preventing cell damage. The key here is that it opens a longer window of time during which surgeons can operate.

Cutting-edge experiments from around 1950 opened the way for DHCA. Using a microscope, Canadian surgeon Wilfred Bigelow revealed unequivocally that cold could protect tissue from a lack of oxygen during circulatory arrest. Cooling human subjects in any circumstance, however, was still highly controversial; it fell under the unyielding stigma created by the discovery of Nazi hypothermia torture during the war. Yet cold's potential was not wasted. In fact, the world's first successful open-heart surgery was performed in 1952 on a five-year-old girl under hypothermic conditions by American cardiac surgeon John Lewis. Lewis had been testing the procedure with dogs.

The success of the surgery spawned further research in England and in Siberia, which showed, incredibly, that although the brain accounts for a mere 2 percent of a person's weight, it guzzles up 20 percent of the blood's oxygen content and receives about the same percentage of the heart's output. And that's just while you're sitting down, at rest. At a normal temperature of about 98.6F° (37°C), permanent brain damage occurs after just six minutes without any circulation. Hypothermia reduces the brain's greedy metabolism by up to 7 percent for every degree Celsius in temperature drop. The metabolism of any cell decreases at an *exponential* rate that equates to about a 50 percent reduction for every 6°C drop in temperature. At 59°F (15°C), the brain's metabolic rate is only 17 percent of normal. Cooling also prevents a host of way-too-complex-for-this-book processes that ultimately lead to cell death. Scientists are still trying to fully understand many of them.

Essentially, DHCA begins when the patient's circulatory system is connected to a heart-lung machine, initiating cardiopulmonary bypass—circulation without the use of the heart and lungs. As the machine circulates and cools the blood, the body and brain are quickly chilled. The patient's head is often additionally cooled to around 57°F (14°C) by a special helmet. When the target temperature is reached, the blood pump is turned off. Circulatory arrest begins. By this time, the brain has gone into electrocerebral silence—brain death.

Doctors then have a time window, the length of which is debated, during which they can perform surgery before the patient risks neural damage. There is some agreement that the magic number is forty minutes, but circulatory arrest has been maintained for up to one hour without serious consequences. Surprisingly, infants generally tolerate longer periods of clinical death under DHCA than adults. Hypothermia is routinely used on newborns as a neuroprotective measure while they undergo treatment for asphyxia, a condition commonly mistaken for stillbirth.

After surgery, the steps taken for cooling are reversed. The patient is given barbiturates to maintain neural inactivity in the brain while they are carefully rewarmed. The heart-lung machine is used for blood circulation before the patient's own heartbeat returns. Often, before reaching a normal rhythm, the heart flutters or beats too quickly. In that case, it is given some help from electrical stimulators. When the patient reaches a normal body temperature, rewarming stops. Any extra warmth causes metabolic demand beyond what can be supplied by the blood. Patients who have been unintentionally overwarmed make up the largest number of cases of disability, coma, and vegetative states.

The procedure has been performed successfully on thousands of patients with a mortality rate ranging between 2 percent and 10 percent, depending on the type of surgery and duration of circulatory arrest.

## EMERGENCY PRESERVATION AND RESUSCITATION

In extreme circumstances, some patients are carefully cooled to temperatures below deep hypothermia's scant 57°F (14°C) threshold—lower even than the coldest temperatures on record that have been survived by victims of accidental hypothermia. We're now talking about what's considered *profound* hypothermia. The success of this treatment is a testament to the evidence and insight gathered by pioneering medics over decades of rigorous scientific research. Even more extraordinarily, the practice of profound hypothermia is reserved for patients who have already suffered a fatal trauma and are clinically dead. Compounding their trauma with cold can actually bring them back to life. Hypothermia as a therapeutic agent simply doesn't get any cooler than this.

In the United States, trauma is the leading cause of death between the ages of one and forty-six. Worldwide it is in the top three causes of fatality for this age group. Although car accidents are the most frequent killer, leading causes of severe trauma include gunshot wounds and stabbings ensuing from senseless, preventable violence.

Victims of severe trauma die all too often from injuries that are actually treatable. Stabbing and gunshot wounds, for example, are frequently the sources of fatal blood loss that happens before surgeons can repair the damage at its source. They simply run out of time.

One surgeon was particularly affected by his experience under such circumstances. As a young trainee, Sam Tisherman, now a professor of surgery at the University of Maryland, was part of a team called in to treat a twenty-three-year-old man who had been stabbed in the heart. The violence had escalated from an argument over bowling shoes.

Although the patient was clearly dying, Tisherman was optimistic about saving him because his injury was treatable with well-established methods. The surgical team knew exactly what to do and how to do it. The patient was intubated and given blood, his chest cavity was opened, his heart was massaged, and his injuries were being repaired. The medical team was close to saving him. But his heart simply wouldn't continue to beat. Despite everything, the patient died. It came down to the simple fact that he had lost too much blood before his heart could be fixed.

The event stayed deeply with Tisherman. He began working at the University of Pittsburgh with Peter Safar, the famous anesthesiologist credited with inventing the lifesaving technique of cardiopulmonary resuscitation (CPR) in the 1950s. As valuable as it is, however, CPR doesn't work on trauma patients. Pumping their chests and helping them breathe has no effect if they are bleeding to death from a gaping wound.

Along with Ron Bellamy, another doctor, Safar and Tisherman began to consider hypothermia as a means of buying time for treating victims' wounds when all other methods of resuscitation fail. They were further inspired after they studied old data from the Vietnam War. After careful analysis, they concluded that soldiers who died within a few hours of sustaining traumatic injury often needlessly met their demise from treatable injuries because there simply wasn't enough time to perform the necessary repairs.

As it does in DHCA, cold can protect a trauma victim's brain; critically, it also protects their heart from the effects of blood loss. These therapeutic properties arise because, as discussed throughout this book, cold has the effect of slowing metabolism—among a host of other complex properties that indirectly aid protection. For these effects to be maximized so that severe trauma injuries can be repaired in time—up to two hours—patients need to be cooled to a core body temperature of around 50°F (10°C). And they need to be cooled as rapidly as possible.

Bellamy, Safar, and Tisherman began to experiment with dogs, cooling them while mimicking the effects of trauma and then slowly rewarming them. By 1996, they had seen enough success to envision an end to the animal experiments. Tisherman, a former vegetarian, was pleased. They had accumulated enough data to publish an article suggesting that profound hypothermia could be applied to victims of severe trauma when all existing methods of resuscitation had failed.

The idea caused a stir in the medical community. Other doctors, as they expected, thought they were crazy. It's surgical dogma that trauma patients need to be kept warm. Hypothermia prevents blood from clotting and contributes to a poisonous buildup of acid in the victim's bloodstream. Worse, these three complications—hypothermia, acidosis, and coagulopathy—known as the "triad of death," mutually compound each other in a continuous feedback loop. Hypothermia also causes shivering and stress, both of which can easily do in someone already suffering from, say, a deadly gunshot, stab, or shrapnel wound. The mortality rate of accidental profound hypothermia victims stands at around 40 percent. How could such a state possibly be used to *treat* severe trauma victims?

Yet mild hypothermia continued to prove itself as an effective means of saving cardiac arrest victims, and deep hypothermia consistently demonstrated therapeutic value for open-heart surgery patients and neurosurgery patients. Additionally, data from cold-water drowning cases, in which survivors experienced rapid drops in core temperature, supported the possibility of surviving circulatory arrest under extreme cold. The key to survival under profound hypothermia appeared to be a combination of rapid cooling and slow rewarming.

Safar and Tisherman called their idea of buying time by using cold *emergency preservation and resuscitation* (EPR). They considered it an extension of existing cooling methods by using profoundly cold temperatures—a difference in degree but not, fundamentally, in kind. The main difference between EPR and other uses of cooling is that trauma victims have already bled to death and have no pulse.

The EPR procedure is to be undertaken if a trauma victim fails to show any sign of life from conventional resuscitation methods after five minutes. The patient's chest is cracked open, and the descending aorta is clamped. Instead of the heart being massaged by hand—a standard resuscitation procedure—the patient is connected to a heart-lung machine that starts exsanguinating them (draining their blood) before pumping them full of an ice-cold saline solution. At a rate of one gallon a minute, it takes roughly twenty minutes to reduce body temperature to a target of 50°F (10°C). At this point, the patient has sustained a fatal injury; they have no pulse, no blood, and a clamped aorta; and they have been clinically dead for nearly a half hour, with no sign of brain activity. Only then can the trauma surgeon begin operating on the injury itself. When trauma surgery is completed, the resanguination process begins, with the heart held at 50°F (10°C) so it won't beat. Next, the cardiac surgeon prepares the patient for a cardiopulmonary bypass. Once that's finished, the patient is finally given blood and their body rewarmed. There's no guarantee that the heart will restart, at least not on its own. In that case, electrical stimulation can trigger a regular pulse.

By 2002 Safar and Tisherman's team had enough data from animal and other studies to warrant starting an EPR trial with humans. They faced another problem: How to obtain patient consent? Unconscious trauma victims, even if not clinically dead on arrival at the hospital, can't read and sign a consent form. And because the decision to induce hypothermia must be made within seconds after death, it is not feasible to contact a patient's family for consent.

Tisherman moved to Baltimore in 2014 to begin a position at the University of Maryland School of Medicine. With a human trial a realistic possibility, his location there couldn't have been more relevant. Baltimore sees some of the most violence in the United States. FBI data from 2017 indicate that the homicide rate there is ten times greater than the US average—worse than that of many of the world's most violent countries. Statistically the city compares with El Salvador, the country with the highest homicide rate. This translates into a need to treat victims. Gunshot and stabbing patients have only a 5 percent chance of survival upon entering the hospital. There couldn't be a better place, unfortunately, to test the EPR procedure. Would these circumstances create a context in which to try the protocol on patients without their consent, if it could save their lives?

In definitive terms: no. In areas with high crime rates, residents are often suspicious of the medical community rather than trusting of it. The divide occurs along racial lines. After a history of gravely unethical studies, mistrust on the part of minorities is warranted. Consider, for example, the Tuskegee Syphilis Experiment, which took place in the mid-1900s. Treatment in the form of penicillin was deliberately withheld from hundreds of black men, who died while scientists studied the disease over a forty-year period.

Tisherman and his colleagues attempted to solve the problem by creating a "No to EPR" opt-out program and informing communities by setting up information desks in shopping malls and other public spaces. The idea was to make EPR a default procedure when it was necessary. To opt out, community members could wear bracelets that said "NO EPR." The team answered questions, distributed pamphlets, and gave interviews on local TV and radio stations. One can imagine the effort they needed to put into community awareness for this strategy to be effective. The proposal was indeed met with skepticism; Baltimore residents accused the researchers of wanting to treat them as lab rats.

Skeptics in the medical community argued that the procedure could leave patients alive but with irreparable brain damage. Without any blood flow to the brain, permanent brain damage can result in just six minutes; the idea of starving the brain for over two hours seemed criminal.

Tisherman's response was simple. EPR would be used only after existing forms of resuscitation had failed and permanent death was imminent. It would be a last-ditch effort for survival. As for brain damage, if there had been a significant risk, it would have been revealed in the previous animal studies. Indeed, one of the key reasons for cooling the body is to prevent brain damage. Refusing to give patients a further chance at survival under circumstances suggesting the possibility of full neurological recovery would be unfortunate.

By 2014, the trial was approved by the Food and Drug Administration. Tisherman's goal was to run EPR on ten patients and compare their outcomes to those of ten patients who had undergone conventional procedures. Getting the ten EPR patients, however, involved surmounting formidable challenges, one of which was logistical rather than technical. The procedure required a sizable trained team consisting of a trauma surgeon, cardiac surgeon, trauma anesthesiologist, cardiac anesthesiologist, perfusionists, trauma resuscitation staff, and other support staff. The chances that all essential team members would be in the right place at the right time—upon the arrival of a fatally injured trauma victim—were certainly not betting odds.

In November 2019, at a symposium in New York, Tisherman revealed that at least one patient had undergone EPR treatment. Multiple media headlines broadcast that "humans are being placed in suspended animation!" It's possible that more patients have had the procedure; when pressed for numbers Tisherman has remained tight-lipped.

The end of the trial will culminate about thirty years of investment for Tisherman. If it works, the procedure could become a game changer in resuscitation. It would provide a further option for saving a life after existing methods have failed. It could also be extended beyond gunshot and stab victims to other trauma patients—those who have suffered otherwise terminal injuries from car accidents, plane crashes, sporting accidents, natural disasters, or any freak accident. EPR can potentially save the life of any trauma patient when conventional methods for restarting the heart without harming the brain can't.

That said, Tisherman hopes that eventually EPR will become outdated. Concurrent with his trial, cutting-edge research continues to look for ways to achieve the time-buying benefits of hypothermia without the risks posed by cold. In the same way that pharmacologically induced torpor could one day prolong spaceflight, drugs that reduce metabolism could perhaps be used to prolong the time available to treat severe trauma victims without a difficult and dangerous requirement for profound hypothermia—or any hypothermia.

# CONCLUSION

Exploring the potential of cold as a means of therapy has consistently been contentious and controversial. Historically, every significant progression has been followed by tragic missteps and abuses, before more innovative methods rebound with new triumphs and advancements.

Cold therapy had an innocent, harmless start, with localized treatments like Egyptian cold compresses and Galenic cold cream. Nobody knew what cold was, physically, or even that it existed on a continuum with heat. But it worked, for thousands of years. Therapeutic use of cold constituted an early step away from using spells and incantations and toward understanding that treatment could be achieved by physical means.

There was a huge leap in progress when heat and cold were conclusively found to be inherently related to the rate of particle motion. When thermometers were invented to actually measure the phenomenon, it became possible to make objective measurements regarding human temperature.

Unfortunately, based on events unlikely to have involved any objective measurement, it was found that cold could be used to subdue patients with psychological disorders. By the seventeenth century the cold-water douche was routinely used to subdue those who were suffering from an "overheated brain." The practice was soon extended to criminals and basically anyone particularly defiant of the social order. Meanwhile, whole-body applications of cold were enjoyed by thousands in spas that grew increasingly popular throughout continental Europe. Nobody cared that those who were popularizing the "cold-water cure" lacked any real idea about how it might have worked; it was trendy and aristocratic, and it made people feel great.

The Enlightenment brought the dawn of the Scientific Revolution; reliance on experimentation, quantification, discovery, and explanation of natural phenomena increased. Natural philosophers like John Hunter established fundamental connections between cold, heat, and the essence of life. The concept of "suspended animation" was born. Apparent but not actual death was a new discovery, as demonstrated by the revival of seemingly lifeless cold-water drowning victims. Hunter surmised that putrefaction was the only true sign of death and that freezing prevented putrefaction. He set out on a quest to discover whether cold could be applied to humans in a manner that would retain the "vital capacity" while suspending life indefinitely. The invention of the microscope and the discovery of minute organisms that could survive freezing, unaged for extended durations, added to the wacky prospect that humans could perhaps live hundreds or even thousands of years, bridging centuries while in a frozen stasis. Unfortunately, over the next centuries, countless animals paid the ultimate price to test this idea.

Early explorers of the science of cold drew connections between life, the nervous system, and electricity. Before the eyes of thrilled audiences, Giovanni Aldini attempted to reanimate cold human corpses in ostentatious displays that were equal measures scientific endeavor and freak show. Cold proved instrumental in forging existential questions about life and death. These were times of wild speculation and cowboy experimentation that ultimately found success in literary fiction, best exemplified by Mary Shelley in her story of Victor Frankenstein's creature.

As modern scientific methods were established, experiments were borne less out of conjecture and more out of hypothesis and evidence. This new approach yielded revolutionary, fundamental discoveries for the therapeutic applications of cold—mainly its ability to slow metabolism and reduce oxygen requirements in human tissue. Armed with physiological evidence, Temple Fay helped cancer patients by reducing their pain and slowing the growth of their tumors. He took the science from the lab bench and applied it to terminally ill patients. Although these events marked cold's greatest triumph up to that time, its greatest defeat lay only months ahead. Nazis tortured prisoners to establish the limits of humans' ability to survive cold, events that stigmatized the use of human subjects for cold research for decades.

The resurgence of cold research came through the use of animals. In a manner suspiciously reminiscent of John Hunter, experiments with rats, hamsters, rabbits, and primates were undertaken with renewed hope of uncovering a mammalian capacity to survive frozen suspended animation by targeting the heart for fast rewarming. Ethically questionable, the experiments failed. It was beginning to seem like the end for cold research. Thankfully, a wider scientific effort, building on the findings of Temple Fay, focused on cold's complex physiological effects. The scientific exploration of cold's ability to sustain cellular life during periods of reduced oxygen paved the way for the introduction of open-heart surgery and neurosurgical techniques, and it revolutionized the science of resuscitation after cardiac arrest. It was cold's comeback.

Still, progress didn't happen without fringe idealism and experimentation. Optiman and cyborgs were conceived in the 1960s as possible cold-tolerant space voyagers, so theoretically modified and augmented that considering them human was questionable. Bob White's experiments testing live brain and head transplants on monkeys combined rigorous science, unprecedented insight, and disturbing, unethical practices. The first decades of cryonics research mirrored earlier centuries' "who knows, let's try it and see" approaches to life extension. Too many times, the only result was putrid, rotting corpses housed in leaky cryopreservation vessels.

Contrast these historical woes with recent advancements in therapeutic hypothermia for treating deadly medical conditions such as stroke, kidney failure, acute liver failure, traumatic brain injury, and hemorrhage. The EPR trial has the potential to revolutionize lifesaving emergency resuscitation. Even cold-water bathing and swimming are making a comeback, with benefits ranging from pain relief to the treatment of depression—all in a manner supported by hard science. A greater understanding of the complex microscopic effects of hypothermia is advancing practices for treating accidental hypothermia victims who lack any vital signs and appear dead but still have the potential for successful resuscitation. The apparently dead have never had it so good.

Still, the most innovative explorations of cold exist on the fringes of acceptable ethics and of pseudoscience. Explorations into the role played by cold in human torpor, as revealed by a deepening understanding of hibernation, are progressing slowly and carefully. Cold is still being studied as a means of enabling everlasting life through cryonics, vitrifixation, and head transplantation. It is being used to explore the nature of consciousness in ongoing projects like AWARE II.

In some ways, the themes surrounding our current explorations into cold, built on the advancements and knowledge accrued over millennia, remain the same as ever. We continue to use cold to explore the nature of our very existence; it remains instrumental in shaping our definitions of life, death, and consciousness. What will the future hold for the science of cold? Where will human needs and desires next steer its potential?

# ACKNOWLEDGMENTS

After writing about being cold for so long, it's time for some warm, crackling, fireside thank-yous to all those who've helped me along the way.

First, I'd like to thank Eva-Stina and everyone else in my family, who have remained ever so curious about the topics I've been writing about and the progress I've been making. Your enthusiasm has been contagious! Although I've hinted about various, seemingly disconnected subjects such as Galvanism, suspended animation, cyborgs, hibernation for space travel, and near-death experiences, you've been so patient despite wanting to know more about how exactly these subjects relate to cold!

I owe many additional thanks to my editor, Colleen Lawrie, and to Kelley Blewster, our copyeditor, who worked magic from my original manuscript! I'd also like to thank Katie Carruthers-Busser for seeing the production through.

Special thanks go to Jeff Shreeve, my literary agent whose creative innovation most notably added structure and organizational appeal to the flurry of topics in this book. And an extra-special thanks to Athena Bryan, who originally blossomed the idea of a book concerning therapeutic hypothermia. Huge thanks, Athena!

# BIBLIOGRAPHY

## CHAPTER I
### Human Temperature

Blagden, Charles. "XII. Experiments and Observations in an Heated Room." *Philosophical Transactions of the Royal Society of London* 65 (1775): 111–123.

Blagden, Charles. "XLVII. Further Experiments and Observations in an Heated Room." *Philosophical Transactions of the Royal Society of London* 65 (1775): 484–494.

Campbell, Kirsten. "How Does Temperature Affect Metabolism?" Sciencing. April 10, 2018. https://sciencing.com/temperature-affect-metabolism-22581.html.

Chan, Casey. "Why the Human Body Temperature Is 98.6 Degrees." *Gizmodo*, January 1, 2011. https://gizmodo.com/why-the-human-body-temperature-is-98-6-degrees-5722520.

"Locomotive Tries Milk Fuel." *Modern Mechanix*. March 1938. Reproduced December 7, 2006. http://blog.modernmechanix.com/locomotive-tries-milk-fuel/#more.

Osilla, Eva V., Jennifer L. Marsidi, and Sandeep Sharma. "Physiology, Temperature Regulation." StatPearls.com. April 30, 2020. https://www.ncbi.nlm.nih.gov/books/NBK507838/?report=reader.

### Hypothermia as a Deadly Condition

Caulaincourt, Armand-Augustin-Louis, Jean Hanoteau, and George Libaire. *With Napoleon in Russia: The Memoirs of General De Caulaincourt, Duke of Vicenza.* New York: W. Morrow and Company, 1935.

Guly, Henry. "History of Accidental Hypothermia." *Resuscitation* 82, no. 1 (2011): 122–125.

Lankford, Harvey V. "Dull Brains and Frozen Feet: A Historical Essay on Cold." *Wilderness and Environmental Medicine* 27, no. 4 (2016): 526–532.

# BIBLIOGRAPHY

Moricheau-Beaupré, P. J. *A Treatise on the Effects and Properties of Cold, with a Sketch, Historical and Medical, of the Russian Campaign.* Edinburgh: Maclachlan Stewart, 1826.

Pitulko, Vladimir V., et al. "Early Human Presence in the Arctic: Evidence from 45,000-Year-Old Mammoth Remains." *Science* 351, no. 6270 (2016): 260–263.

Pitulko, Vladimir V., et al. "'They Came from the Ends of the Earth': Long-Distance Exchange of Obsidian in the High Arctic During the Early Holocene." *Antiquity* 93, no. 367 (2019): 28–44.

Sullivan, Walter. "The South Pole Fifty Years After." *Arctic* 15, no. 3 (1962): 174–178.

## Miracle in Ice

Gilbert, M., et al. "Resuscitation from Accidental Hypothermia of 13.7 Degrees C with Circulatory Arrest." *Lancet* 355, no. 9201 (2000): 375–376.

Martin, David S. "From an Icy Slope, a Medical Miracle Emerges." CNN.com. 2009. www.cnn.com/2009/HEALTH/10/12/cheating.death.bagenholm/.

## Hypothermia as a Therapeutic Condition

Bohl, Michael A., et al. "The History of Therapeutic Hypothermia and Its Use in Neurosurgery." *Journal of Neurosurgery* 130, no. 3 (2018): 1006–1020.

Breasted, James H. The Edwin Smith Surgical Papyrus. Chicago: University of Chicago Press, 1930.

## Hippocrates Finds the Humor in Cold

Hippocrates, G. E. R. Lloyd, John Chadwick, and W. N. Mann. *Hippocratic Writings.* Harmondsworth, UK: Penguin, 1983.

## Avicenna and the Chilly "Stupefacient"

Avicenna. *The Canon of Medicine of Avicenna.* Edited and translated by O. Cameron Gruner. New York: AMS Press, 1973.

Smith, R. D. "Avicenna and the Canon of Medicine: A Millennial Tribute." *Western Journal of Medicine* 133, no. 4 (1980): 367–370.

# CHAPTER 2

## Heat Exchange:
## From Magic to Measurable Phenomenon

Bigotti, Fabrizio. "The Weight of the Air: Santorio's Thermometers and the Early History of Medical Quantification Reconsidered." *Journal of Early Modern Studies* 7, no. 1 (2018): 73–103.

Camuffo, Dario, and Chiara Bertolin. "The Earliest Spirit-in-Glass Thermometer and a Comparison Between the Earliest CET and Italian Observations." *Weather* 67, no. 8 (2012): 206.

de Grijs, Richard, and Daniel Vuillermin. "Measure of the Heart: Santorio Santorio and the Pulsilogium." *Hektoen International* 9, no. 3 (2017). https://hekint .org/2017/04/17/measure-of-the-heart-santorio-santorio-and-the-pulsilogium/.

Ergonul, Onder, et al. "An Unexpected Tetanus Case." *The Lancet: Infectious Diseases* 16, no. 6 (2016): 746–752. doi:10.1016/S1473-3099(16)00075-X.

Grigull, Ulrich. "Fahrenheit a Pioneer of Exact Thermometry." International Heat Transfer Conference Digital Library. 1986. Accessed March 2020. http://ihtc digitallibrary.com/download/article/1480b040272d93c1/pl-2.pdf.

Guly, Henry. "History of Accidental Hypothermia." *Resuscitation* 82, no. 1 (2011): 122–125.

Haller, J. S., Jr. "Medical Thermometry: A Short History." *Western Journal of Medicine* 142, no. 1 (1985): 108–116.

Lienhard, John H. "No. 45: Fahrenheit." *Engines of Our Ingenuity*. Accessed October 21, 2020. https://uh.edu/engines/epi45.htm.

"Who Invented the Thermometer?" Brannan Thermometers and Instrumentation. 2016. www.brannan.co.uk/who-invented-the-thermometer.

## *Hydro*therapy?

Claridge, R. T. *Hydropathy; or, the Cold Water Cure, as Practised by Vincent Priessnitz, at Graefenberg, Silesia, Austria.* London: James Madden and Co., 1843.

Currie, J. *Medical Reports, on the Effects of Water, Cold and Warm: As a Remedy in Fever and Other Diseases, Whether Applied to the Surface of the Body, or Used Internally.* Vol. 1. London: T. Cadell and W. Davies, 1805.

Floyer, John. *Psycholusia; or, the History of Cold-Bathing, Both Ancient and Modern.* London: W. Innys, 1732.

Forrester, John M. "The Origins and Fate of James Currie's Cold Water Treatment for Fever." *Medical History* 44, no. 1 (2000): 57–74.

Lindsay, L. "Sir John Floyer (1649–1734)." *Proceedings of the Royal Society of Medicine* 44, no. 1 (1951): 43–48.

Locher, Cornelia, and Christof Pforr. "The Legacy of Sebastian Kneipp: Linking Wellness, Naturopathic, and Allopathic Medicine." *Journal of Alternative and Complementary Medicine* 20, no. 7 (2014): 521–526.

Shapiro, R. W. "James Currie: The Physician and the Quest." *Medical History* 7, no. 3 (1963): 212–231.

## *Cold-Water Swimming*

Tipton, Mike. "Is a Cold Water Swim Good for You, or More Likely to Send You to the Bottom?" *The Conversation.* January 1, 2018. https://theconversation .com/is-a-cold-water-swim-good-for-you-or-more-likely-to-send-you-to-the -bottom-89513.

Tipton, Mike, et al. "Cold Water Immersion: Kill or Cure?" *Experimental Physiology* 102, no. 11 (2017): 1335–1355.

## *Cooling and Pseudopsychology*

Braslow, Joel. *Mental Ills and Bodily Cures: Psychiatric Treatment in the First Half of the Twentieth Century.* Vol. 8. Berkeley: University of California Press, 1997.

Cox, Stephanie C., et al. "Showers: From a Violent Treatment to an Agent of Cleansing." *History of Psychiatry* 30, no. 1 (2019): 58–76.

Foucault, Michel. *Madness and Civilization: A History of Insanity in the Age of Reason.* New York: Vintage, 1988.

Helmont, Franciscus M. *The Spirit of Diseases; or, Diseases from the Spirit: Laid Open in Some Observations Concerning Man and His Diseases.* London: Printed for Sarah Howkins, 1694.

International Committee of the Red Cross. *ICRC Report on the Treatment of Fourteen "High Value Detainees" in CIA Custody.* Website of *New York Review of Books.* February 14, 2007. http://www.nybooks.com/media/doc/2010/04/22/icrc-report.pdf.

Morison, Alexander. *Cases of Mental Disease with Practical Observations on the Medical Treatment.* London: Longman, 1828.

Rowley, William. *A Treatise on Female Nervous Hysterical, Hypochondriacal, Bilious, Convulsive Diseases.* London: C. Nourse, 1788.

Zhang, Sarah. "Showering Has a Dark, Violent History." *The Atlantic,* December 11, 2018. www.theatlantic.com/health/archive/2018/12/dark-history-of-showering/577636/.

# CHAPTER 3

## *Temple Fay: Crucial Advancements, Angry Nurses*

Alzaga, Ana G., et al. "Resuscitation Great: Breaking the Thermal Barrier: Dr. Temple Fay." *Resuscitation* 69, no. 3 (2006): 359–364.

Bohl, Michael A., et al. "The History of Therapeutic Hypothermia and Its Use in Neurosurgery." *Journal of Neurosurgery* 130, no. 3 (2018): 1–15.

Bruen, Charles. "Spontaneous Circulation: A History of General Refrigeration." *Emergency Medicine News* 37, no. 4 (2015): 14–15.

## *Disaster Strikes*

Alexander, L. "Medical Science Under Dictatorship." *New England Journal of Medicine* 241, no. 2 (1949): 39–47.

Berger, R. L. "Ethics in Scientific Communication: Study of a Problem Case." *Journal of Medical Ethics* 20, no. 4 (1994): 207–211.

# BIBLIOGRAPHY

Berger, R. L. "Nazi Science: The Dachau Hypothermia Experiments." *New England Journal of Medicine* 322, no. 20 (1990): 1435–1440.

Bogod, David. "The Nazi Hypothermia Experiments: Forbidden Data?" *Anaesthesia* 59, no. 12 (2004): 1155–1156.

Caplan, Arthur L. *When Medicine Went Mad: Bioethics and the Holocaust*. New York: Springer Science and Business Media, 2012.

Proctor, R. N. "Nazi Science and Nazi Medical Ethics: Some Myths and Misconceptions." *Perspectives in Biology and Medicine* 43, no. 3 (2000): 335–346.

Schafer, Arthur. "On Using Nazi Data: The Case Against." *Dialogue* 25, no. 3 (1986): 413–419.

Siegel, Barry. "Can Evil Beget Good?: Nazi Data: A Dilemma for Science." *Los Angeles Times*, October 30, 1988. www.latimes.com/archives/la-xpm-1988-10-30-mn-958-story.html.

Steinberg, Jonathan. "The Ethical Use of Unethical Human Research." *Rutgers Journal of Bioethics* 5 (2014): 8–15.

Wilson, Sarah. "The Nazi Research Data: Should We Use It?" *CedarEthics: A Journal of Critical Thinking in Bioethics* 10, no. 2 (2011): 1–6.

## *"Refrigeration Therapy"*

de Young, Mary. *Encyclopedia of Asylum Therapeutics, 1750–1950s*. Jefferson, NC: McFarland, 2015.

Dill, D. B., and W. H. Forbes. "Respiratory and Metabolic Effects of Hypothermia." *American Journal of Physiology: Legacy Content* 132, no. 3 (1941): 685–697.

Hoen, T. I., et al. "Hypothermia (Cold Narcosis) in the Treatment of Schizophrenia." *Psychiatric Quarterly* 31, no. 4 (1957): 696–702.

Lagnado, Lucette. "A Scientist's Nazi-Era Past Haunts Prestigious Space Prize." *Wall Street Journal*, December 1, 2012. www.wsj.com/articles/SB10001424052970204349404578101393870218834.

Newman, Barclay Moon. "Refrigeration for Insanity." *Scientific American* 167, no. 4 (1942): 160–162.

Talbott, John H., W. V. Consolazio, and L. J. Pecora. "Hypothermia: Report of a Case in Which the Patient Died During Therapeutic Reduction of Body Temperature, with Metabolic and Pathologic Studies." *Archives of Internal Medicine* 68, no. 6 (1941): 1120–1132.

Talbott, J. H., and K. J. Tillotson. "The Effects of Cold on Mental Disorders: A Study of Ten Patients Suffering from Schizophrenia and Treated with Hypothermia." *Diseases of the Nervous System* 2 (1941): 116–126.

Véghelyi, Peter V. "Artificial Hibernation." *Journal of Pediatrics* 60, no. 1 (1962): 122–138.

Whitaker, Robert. *Mad in America: Bad Science, Bad Medicine, and the Enduring Mistreatment of the Mentally Ill*. New York: Basic Books, 2001.

# BIBLIOGRAPHY

## Eureka!

Crossman, Lyman Weeks, et al. "Reduced Temperatures in Surgery, II: Amputations for Peripheral Vascular Disease." *Archives of Surgery* 44, no. 1 (1942): 139–156.

Crossman, Lyman Weeks, and Frederick M. Allen. "Shock and Refrigeration." *Journal of the American Medical Association* 130, no. 4 (1946): 185–189.

Davidson, Armstrong. "The Evolution of Anaesthesia." *British Journal of Anaesthesia* 31, no. 9 (1959): 417–422.

Griffin, Alexander R. *Out of Carnage*. Auckland, NZ: Pickle Partners Publishing, 2019.

"Ice Anaesthesia." *Life*, April 22, 1946.

Rosomoff, Hubert L. "Historical Review of the Development of Brain Hypothermia." In *Hypothermia for Acute Brain Damage*, edited by N. Hayashi et al., 3–16. Tokyo: Springer, 2004.

# CHAPTER 4
## Legendary Hibernators

"Human Hibernation." *British Medical Journal* (Clinical Research Ed.) 320, no. 7244 (2000 [1900]): 1245A.

## Lady Sleepers

Grossman, Ron. "Chicago's 'Frozen Woman': The Amazing, True Story of Dorothy Mae Stevens." *Chicago Tribune*, January 24, 2019. www.chicagotribune.com /opinion/commentary/ct-perspec-flashback-frozen-woman-dorothy-mae -stevens-cold-weather-20190129-story.html.

Madea, Burkhard, and Eberhard Lignitz. "Die lebende Tote und andere Fehlleistungen bei der ärztlichen Leichenschau" [The Living Dead and Other Failures of Medical Examination]. In *Von den Maden zum Mörder: Die vielfältigen Ermittlungsmethoden der Rechtsmedizin* [From Maggots to Murder: The Diverse Investigative Methods of Forensic Medicine], edited by Burkhard Madea. Leipzig: Militzke Verlag, 2011.

## The Space Race

Bimm, Jordan, and Patrick Kilian. "The Well-Tempered Astronaut." *Nach Feierabend: Der Kalte Krieg* (2017): 85–107.

Lagnado, Lucette. "A Scientist's Nazi-Era Past Haunts Prestigious Space Prize." *Wall Street Journal*, December 1, 2012. www.wsj.com/articles/SB10001424052970204 349404578101393870218834.

## *"Optiman" or Cyborg?*

Freedman, Toby, and Gerald S. Linder. "Optiman! For the Space Age." *Space Digest* 46 (1963): 43–46.

Kline, Nathan S., and Manfred Clynes. "Drugs, Space, and Cybernetics: Evolution to Cyborgs." In *Psychophysiological Aspects of Space Flight*, edited by Bernard E. Flaherty, 345–371. New York: Columbia University Press, 1961.

Rosenfeld, A. "The Last Barrier Is Man Himself." *Life*, November 30, 1964.

## *Hibernation and Torpor for Prolonged Space Travel*

Bradford, John, M. Schaffer, and D. Talk. "Torpor Inducing Transfer Habitat for Human Stasis to Mars." *NASA Innovative Advanced Concepts (NIAC) Phase I Final Report*. May 11, 2014. www.nasa.gov/sites/default/files/files/Bradford_2013 _PhI_Torpor.pdf.

Cockett, A. T., and C. C. Beehler. "Total Body Hypothermia for Prolonged Space Travel." *Aerospace Medicine* 34 (1963): 504.

Gemignani, Jessica, Tom Gheysens, and Leopold Summerer. "Beyond Astronaut's Capabilities: The Current State of the Art." *37th Annual International Conference of the IEEE Engineering in Medicine and Biology Society (EMBC)* (2015): 3615–3618. https://ieeexplore.ieee.org/document/7319175.

Jackson, Travis C., and Patrick M. Kochanek. "A New Vision for Therapeutic Hypothermia in the Era of Targeted Temperature Management: A Speculative Synthesis." *Therapeutic Hypothermia and Temperature Management* 9, no. 1 (2019): 13–47.

Jiang, Ji-Yao, Ming-Kun Yu, and Cheng Zhu. "Effect of Long-Term Mild Hypothermia Therapy in Patients with Severe Traumatic Brain Injury: 1-Year Follow-Up Review of 87 Cases." *Journal of Neurosurgery* 93, no. 4 (2000): 546–549.

McCurry, Justin, and Alok Jha. "Injured Hiker Survived 24 Days on Mountain by 'Hibernating.'" *The Guardian*, December 21, 2006. www.theguardian.com /world/2006/dec/21/japan.topstories3.

Ng, H. Kee, R. A. Hanel, and W. D. Freeman. "Prolonged Mild-to-Moderate Hypothermia for Refractory Intracranial Hypertension." *Journal of Vascular and Interventional Neurology* 2, no. 1 (2009): 142.

Strughold, H. "Atmospheric Space Equivalence." *Journal of Aviation Medicine* 25, no. 4 (1954): 420.

Whillans, M. G. "Symposium on Space-Biosciences Research and Space Problems." *Journal of the Royal Astronomical Society of Canada* 54 (1960): 211.

## *Hibernation or Torpor?*

Bradford, Schaffer, and Talk. "Torpor Inducing Transfer Habitat for Human Stasis to Mars."

Choukèr, A., et al. "Hibernating Astronauts: Science or Fiction?" *Pflügers Archiv - European Journal of Physiology* 471, no. 6 (2019): 819–828.

Nordeen, Claire A., and Sandra L. Martin. "Engineering Human Stasis for Long-Duration Spaceflight." *Physiology* 34, no. 2 (2019): 101–111.

# CHAPTER 5

## *Frozen to Life*

Bohl, Michael A., et al. "The History of Therapeutic Hypothermia and Its Use in Neurosurgery." *Journal of Neurosurgery* 130, no. 3 (2018): 1006–1020.

Breathnach, Caoimhghin S., and John B. Moynihan. "Intensive Care 1650: The Revival of Anne Greene (c. 1628–59)." *Journal of Medical Biography* 17, no. 1 (2009): 35–38.

Keilin, David. "The Leeuwenhoek Lecture: The Problem of Anabiosis or Latent Life: History and Current Concept." *Proceedings of the Royal Society of London, Series B: Biological Sciences* 150, no. 939 (1959): 149–191.

Solly, Meilan. "Ancient Roundworms Allegedly Resurrected from Russian Permafrost." *Smithsonian*, July 30, 2018. www.smithsonianmag.com/smart-news/ancient-roundworms-allegedly-resurrected-russian-permafrost-180969782/.

Watkins, Richard. *Newes from the Dead; or, a True and Exact Narration of the Miraculous Deliverance of Anne Greene.* London: L. Lichfield, 1651.

## *Mad Science*

Baker, Henry. *Employment for the Microscope in Two Parts: I. An Examination of Salts and Saline Substances; II. An Account of Various Animalcules; Likewise, a Description of the Microscope.* London: R. Dodsley, 1753.

Boyle, Robert. *New Experiments and Observations Touching Cold, or, an Experimental History of Cold Begun to Which Are Added an Examen of Antiperistasis and an Examen of Mr. Hobs's Doctrine About Cold by the Honorable Robert Boyle.* London: John Crook, 1665.

Boyle, Robert. *Philosophical Works of the Honourable Robert Boyle.* London: W. and J. Innys, 1725.

Wharton, David A. *Life at the Limits: Organisms in Extreme Environments.* Cambridge, UK: Cambridge University Press, 2007.

## *John Hunter's Secret Plan*

Bohl, Michael A., et al. "The History of Therapeutic Hypothermia and Its Use in Neurosurgery." *Journal of Neurosurgery* 130, no. 3 (2018): 1006–1020.

Bud, Robert, Bernard S. Finn, and Helmuth Trischler, eds. *Manifesting Medicine: Bodies and Machines.* Vol. 1. Amsteldijk, NL: Harwood, 1999.

Guthrie, Douglas. "John Hunter: Surgeon and Naturalist." *Edinburgh Medical Journal* 49, no. 2 (1942): 119.

Hunter, John. "XXIV. Proposals for the Recovery of People Apparently Drowned." *Philosophical Transactions of the Royal Society of London* 66 (1776): 412–425.

Hunter, John. *Lectures on the Principles of Surgery*. London: Haswell, Barrington, and Haswell, 1839.

Hunter, John, and James F. Palmer. *The Works of John Hunter: With Notes; in Four Volumes*. London: Longman, Rees, Orme, Brown, Green and Longman, 1837.

Kapp, K., and G. Talboy. *John Hunter: The Father of Scientific Surgery*. American College of Surgeons, Bulletin of the Surgery History Group, 2017. www.facs.org /~/media/files/archives/shg%20poster/2017/05_john_hunter.ashx.

Mitchell, Robert. "Suspended Animation, Slow Time, and the Poetics of Trance." *Publications of the Modern Language Association* 126, no. 1 (2011): 107–122.

Reill, Peter H. *Vitalizing Nature in the Enlightenment*. Los Angeles: University of California Press, 2005.

Wakeley, Cecil. "John Hunter and Experimental Surgery: Hunterian Oration Delivered at the Royal College of Surgeons of England on 14th February 1955." *Annals of the Royal College of Surgeons of England* 16, no. 2 (1955): 69.

## *Nervous Breakdown*

Aldini, Giovanni. *An Account of the Late Improvements in Galvanism; with a Series of Curious and Interesting Experiments . . . to Which Is Added an Appendix Containing Experiments on the Body of a Malefactor Executed at Newgate, Etc.* London: Cuthell and Martin and J. Murray, 1803.

Bohl, Michael A., et al. "The History of Therapeutic Hypothermia and Its Use in Neurosurgery." *Journal of Neurosurgery* 130, no. 3 (2018): 1006–020

Cajavilca, Christian, Joseph Varon, and George L. Sternbach. "Luigi Galvani and the Foundations of Electrophysiology." *Resuscitation* 80, no. 2 (2009): 159–162.

Cambiaghi, M., and A. Parent. "From Aldini's Galvanization of Human Bodies to the Modern Prometheus." *Medicina Historica* 2, no. 1 (2018): 27–37.

Hurren, Elizabeth. "Bibliography: Staging Post-Execution Punishment in Early Modern England." *Dissecting the Criminal Corpse*. New York: Springer Nature, 2016.

Pernick, Martin S. "Back from the Grave: Recurring Controversies over Defining and Diagnosing Death in History." *In Death: Beyond Whole-Brain Criteria*, edited by Richard N. Zaner, 17–74. Dordrecht, NL: Springer, 1988.

"William Cullen Biography: Inventor of Artificial Refrigeration." History of Refrigeration. Accessed April 2020. www.historyofrefrigeration.com/refrigeration-invention/william-cullen/.

## *The End of an Era?*

Andjus, Pavle R., Stanko S. Stojilkovic, and Gordana Cvijic. "Ivan Djaja (Jean Giaja) and the Belgrade School of Physiology." *Physiological Research/Academia Scientiarum Bohemoslovaca* 60, suppl. 1 (2011): S1.

# BIBLIOGRAPHY

Andjus, R. K., J. E. Lovelock, and A. U. Smith. "Resuscitation and Recovery of Hypo-thermic, Supercooled and Frozen Mammals." *Proceedings of the National Academy of Sciences* 451 (1956): 125.

Castle, Matt. "Reanimated Rodents and the Meaning of Life." Damn Interesting. December 5, 2012. www.damninteresting.com/reanimated-rodents-and -the-meaning-of-life/.

Keilin, David. "The Leeuwenhoek Lecture: The Problem of Anabiosis or Latent Life: History and Current Concept." *Proceedings of the Royal Society of London, Series B: Biological Sciences* 150, no. 939 (1959): 149–191.

Lovelock, James. *Homage to Gaia: The Life of an Independent Scientist.* New York: Oxford University Press, 2000.

Parkes, A. S., and Audrey U. Smith. "Transport of Life in the Frozen or Dried State." *British Medical Journal* 1, no. 5132 (1959): 1295.

Parry, Bronwyn. "Technologies of Immortality: The Brain on Ice." *Studies in History and Philosophy of Science Part C: Studies in History and Philosophy of Biological and Biomedical Sciences* 35, no. 2 (2004): 391–413.

Smith, Audrey. "Resuscitation of Frozen Mammals." *New Scientist* 4, no. 102 (1958): 1153–1155.

Smith, Audrey U. "Problems in the Resuscitation of Mammals from Body Temperatures Below 0°C." *Proceedings of the Royal Society of London, Series B: Biological Sciences* 147, no. 929 (1957): 533–544.

## CHAPTER 6

Best, Ben. "A History of Cryonics." Website of Ben Best. 2018. www.benbest.com /cryonics/history.html.

Glass, Ira. "Mistakes Were Made." *This American Life.* April 18, 2008.

Moen, Ole Martin. "The Case for Cryonics." *Journal of Medical Ethics* 41, no. 8 (2015): 677–681.

Nelson, Bob, Kenneth Bly, and Sally Magana. *Freezing People Is (Not) Easy: My Adventures in Cryonics.* Lanham, MD: Rowman and Littlefield, 2014.

Parry, Bronwyn. "Technologies of Immortality: The Brain on Ice." *Studies in History and Philosophy of Science Part C: Studies in History and Philosophy of Biological and Biomedical Sciences* 35, no. 2 (2004): 391–413.

Perry, Michael. "First Suspension No 'Blue Sky' Event." *Cryonics* 12 (1991): 11–14.

Perry, Michael. "Suspension Failures: The Dark Side of Cryonics History." *Cryonics* 13 (1991): 5–8.

Smith, Audrey. "Resuscitation of Frozen Mammals." *New Scientist* 4, no. 102 (1958): 1153–1155.

Starr, Michelle. "Cool Dude James Bedford Has Been Cryonically Frozen for 50 Years." *CNET,* January 12, 2017. www.cnet.com/news/cool-dude-james -bedford-has-been-cryonically-frozen-for-50-years/.

Styles, Ruth. "Cryogenics Pioneer Speaks Out About His First 'Patient.'" *Daily Mail Online,* January 30, 2017. www.dailymail.co.uk/news/article-4156504/Pioneer -cryogenics-speaks-patient.html.

Vitello, Paul. "Robert C. W. Ettinger, a Proponent of Life After (Deep-Frozen) Death, Is Dead at 92." *New York Times*, July 30, 2011. www.nytimes.com/2011/07/30 /us/30ettinger.html.

Walker, David. "Valley Cryonic Crypt Desecrated, Untended." *Valley News*, June 10, 1979.

## *A New Beginning*

Darwin, Michael. "The Dog and Pony Show." *Cryonics* 8 (1987): 3–9.

Darwin, Michael. "Jerry Leaf Enters Cryonic Suspension." *Cryonics* 12 (1991): 19–26.

Hayworth, Ken. "Opinion: The Prize Win Is a Vindication of the Idea of Cryonics, Not of Unaccountable Cryonics Service Organizations." Brain Preservation Foundation. February 9, 2016. www.brainpreservation.org/opinion-the-prize -win-is-a-vindication-of-the-idea-of-cryonics-not-of-unaccountable-cryonics -service-organizations/.

Perry, Michael. "Our Finest Hours: Notes on the Dora Kent Crisis (Part 1)." *Cryonics* 13, no. 9 (1992): 4–6.

Perry, Michael. "Our Finest Hours: Notes on the Dora Kent Crisis (Part 2)." *Cryonics* 13, no. 10 (1992): 4–6.

Perry, Michael. "Our Finest Hours: Notes on the Dora Kent Crisis (Part 3)." *Cryonics* 13, no. 11 (1992): 4–8.

Swan, Melanie. "Worldwide Cryonics Attitudes About the Body, Cryopreservation, and Revival: Personal Identity Malleability and a Theory of Cryonic Life Extension." *Sophia* 58, no. 4 (2019): 699–735.

Warren, Luigi. "Interview with Michael Darwin." *Cryonics* 7 (1986): 23–33.

## *Cryonics Matures*

Society for Cryobiology. "Position Statement: Cryonics." November 2018. https:// www.societyforcryobiology.org/assets/documents/Position_Statement _Cryonics_Nov_18.pdf.

# CHAPTER 7

Webb, Adam C., and Owen B. Samuels. "Reversible Brain Death After Cardiopulmonary Arrest and Induced Hypothermia." *Critical Care Medicine* 39, no. 6 (2011): 1538–1542.

## *Traveling Without Moving*

Carter, Chris. "Response to 'Could Pam Reynolds Hear?'" *Journal of Near-Death Studies* 30, no. 1 (2011): 29–53.

Greyson, Bruce. "The Near-Death Experience Scale." *Journal of Nervous and Mental Disease* 171, no. 6 (1983): 369–375.

Hagerty, Barbara Bradley. "Decoding the Mystery of Near-Death Experiences." *NPR*, May 22, 2009. www.npr.org/templates/story/story.php?storyId=104397005.

Kirkey, Sharon. "Life—After Life: Does Consciousness Continue After Our Brain Dies?" *National Post*, April 18, 2019. nationalpost.com/news/canada/life-after -life-does-consciousness-continue-after-our-brain-dies.

Marsh, Michael N. "The Near-Death Experience: A Reality Check?" *Humanities* 5, no. 2 (2016): 18.

Parnia, Sam. "Understanding the Cognitive Experience of Death and the Near-Death Experience." *QJM: An International Journal of Medicine* 110, no. 2 (2017): 67–69.

Parnia, Sam, et al. "AWARE—AWAreness During REsuscitation: A Prospective Study." *Resuscitation* 85, no. 12 (2014): 1799–1805.

Parnia, Sam, et al. "Awareness and Cognitive Activity During Cardiac Arrest." *Circulation* 140, suppl. 2 (2019): A387.

Sharp, Kimberly Clark. "The Other Shoe Drops: Commentary on 'Does Paranormal Perception Occur in Near-Death Experiences?'" *Journal of Near-Death Studies* 25, no. 4 (2007): 245–250.

Wehrstein, Karen. "Pam Reynolds (Near-Death Experience)." Psi Encyclopedia. October 20, 2017. psi-encyclopedia.spr.ac.uk/articles/pam-reynolds-near-death -experience.

Woerlee, Gerald M. "Could Pam Reynolds Hear? A New Investigation into the Possibility of Hearing During This Famous Near-Death Experience." *Journal of Near-Death Studies* 30, no. 1 (2011): 3–25.

## *Brain Freeze*

Ausman, James I. "Is It Time to Perform the First Human Head Transplant? Comment on the CSA (Cephalosomatic Anastomosis) Paper by Ren, Canavero, and Colleagues." *Surgical Neurology International* 9 (2018). doi:10.4103/sni.sni _471_17.

Berko, Lex. "Meet the Late Dr. Robert White, Who Transplanted the First Monkey Head." *Vice*, July 11, 2013. www.vice.com/en_us/article/pggnk7/dr-robert -white-transplanted-first-monkey-head.

Bohl, Michael A., et al. "The History of Therapeutic Hypothermia and Its Use in Neurosurgery." *Journal of Neurosurgery* 130, no. 3 (2018): 1006–1020.

Canavero, Sergio. "HEAVEN: The Head Anastomosis Venture Project Outline for the First Human Head Transplantation with Spinal Linkage (GEMINI)." *Surgical Neurology International* 4, suppl. 1 (2013): S335.

Furr, Allen, et al. "Surgical, Ethical, and Psychosocial Considerations in Human Head Transplantation." *International Journal of Surgery* 41 (2017): 190–195.

Gkasdaris, Grigorios, and Theodossios Birbilis. "First Human Head Transplantation: Surgically Challenging, Ethically Controversial and Historically Tempting—An Experimental Endeavor or a Scientific Landmark?" *Maedica* 14, no. 1 (2019): 5.

Kean, Sam. "The Audacious Plan to Save This Man's Life by Transplanting His Head." *The Atlantic*, September 2016. www.theatlantic.com/magazine/archive/2016/09 /the-audacious-plan-to-save-this-mans-life-by-transplanting-his-head/492755/.

Kirkey, Sharon. "First Man to Sign Up for Head Transplant Bows Out, but Surgeon Insists List of Volunteers Is Still 'Quite Long.'" *National Post*, April 9, 2019. nationalpost.com/news/world/first-man-to-sign-up-for-head-transplant-bows -out-but-surgeon-insists-list-of-volunteers-is-still-quite-long.

Kirkey, Sharon. "Meet the Man Proposing the First Brain Transplant and, Perhaps, Everlasting Life." *National Post*, December 10, 2016.

Kleban, Barbara. "A Devout Neurosurgeon Studies the Brain with Medically Dazzling, Morally Puzzling Head Transplants." *People*, August 13, 1979.

Lamba, Nayan, Daniel Holsgrove, and Marike L. Broekman. "The History of Head Transplantation: A Review." *Acta Neurochirurgica* 158, no. 12 (2016): 2239–2247.

Lang, Min, et al. "A Tribute to Dr. Robert J. White." *Neurosurgery* 85, no. 2 (2019): E366–E373.

Malchesky, Paul S. "Robert J. White: Renowned Neurosurgeon, Author, Bioethicist, Medical Adviser to Popes. A Man for All Seasons." *Artificial Organs* 35, no. 1 (2011): 1–3.

Manjila, Sunil, et al. "From Hypothermia to Cephalosomatic Anastomoses: The Legacy of Robert White (1926–2010) at Case Western Reserve University of Cleveland." *World Neurosurgery* 113 (2018): 14–25.

Segall, Grant. "Dr. Robert J. White, Famous Neurosurgeron and Ethicist, Dies at 84." *Cleveland.com*, September 16, 2010. www.cleveland.com/obituaries/2010/09 /dr_robert_j_white_was_a_world-.html.

White, Robert J., et al. "Prolonged Whole-Brain Refrigeration with Electrical and Metabolic Recovery." *Nature* 209, no. 5030 (1966): 1320–1322.

## *"Vitrifixation"*

Begley, Sharon. "After Ghoulish Allegations, a Brain-Preservation Company Seeks Redemption." *STAT*, January 30, 2019. www.statnews.com/2019/01/30/nectome -brain-preservation-redemption/.

Berman, Robby. "Startup Offers a Path to Immortality. The Catch? It's '100% Fatal.'" Big Think. April 26, 2019. bigthink.com/robby-berman/company-offers-a -killer-new-way-to-upload-your-mind.

McIntyre, Robert L., and Gregory M. Fahy. "Aldehyde-Stabilized Cryopreservation." *Cryobiology* 71, no. 3 (2015): 448–458.

O'Brien, Kelly J. "MIT Cuts Links to Startup Working on Fatal Brain-Uploading Tech." *Bizjournals*, April 3, 2018. www.bizjournals.com/boston/news/2018/04/03 /mit-cuts-links-to-startup-working-on-fatal-brain.html.

Prisco, Giulio. "I Want to Preserve My Brain So My Mind Can Be Uploaded to a Computer in the Future." *Vice*, April 2, 2018. www.vice.com/en_us/article /43baam/uploading-the-mind-to-a-computer-cryonics.

Regalado, Antonio. "A Startup Is Pitching a Mind-Uploading Service That Is '100 Percent Fatal.'" *MIT Technology Review*, March 13, 2018. www.technology review.com/2018/03/13/144721/a-startup-is-pitching-a-mind-uploading -service-that-is-100-percent-fatal/.

Wiley, Keith. "Implications of the BPF Large Mammal Brain Preservation Prize." Brain Preservation Foundation. March 13, 2018. www.brainpreservation.org /implications-of-the-bpf-large-mammal-brain-preservation-prize/.

# CHAPTER 8
## Dr. Freeze

"Doctor Uses Cooling Treatment to Save His Own Life." ABC13 Eyewitness News. December 8, 2014. abc13.com/dr-varon-stroke-physician-patient/426432/.

Fong, Petti. "The Iced Man Liveth: Miracle Helps Canadian Cheat Death." *Globe and Mail*, March 8, 2005. www.theglobeandmail.com/life/the-iced-man-liveth -miracle-helps-canadian-cheat-death/article4115968/.

Gillis, Charlie. "Lazarus on Ice." *Maclean's*, April 11, 2005. archive.macleans.ca /article/2005/4/11/lazarus-on-ice.

### Current Contentions

Bernard, Stephen A., et al. "Treatment of Comatose Survivors of Out-of-Hospital Cardiac Arrest with Induced Hypothermia." *New England Journal of Medicine* 346, no. 8 (2002): 557–563.

Bohl, Michael A., et al. "The History of Therapeutic Hypothermia and Its Use in Neurosurgery." *Journal of Neurosurgery* 130, no. 3 (2018): 1006–1020.

Corry, Jesse J., et al. "Hypothermia for Refractory Status Epilepticus." *Neurocritical Care* 9, no. 2 (2008): 189.

Giesinger, Regan E., et al. "Hypoxic-Ischemic Encephalopathy and Therapeutic Hypothermia: The Hemodynamic Perspective." *Journal of Pediatrics* 180 (2017): 22–30.

Hasslacher, Julia, et al. "Acute Kidney Injury and Mild Therapeutic Hypothermia in Patients After Cardiopulmonary Resuscitation: A Post Hoc Analysis of a Prospective Observational Trial." *Critical Care* 22, no. 1 (2018): 154.

Helmy, Adel, Hakim Abdelkhellaf, and Danyal Khan. "Recent Advances in Traumatic Brain Injury." *Journal of Neurology* 266, no. 11 (2019): 2878–2889.

Kapinos, Gregory, and Lance B. Becker. "The American Academy of Neurology Affirms the Revival of Cooling for the Revived." *Neurology* 88, no. 22 (2017): 2076–2077.

Mohsenin, Vahid. "Assessment and Management of Cerebral Edema and Intracranial Hypertension in Acute Liver Failure." *Journal of Critical Care* 28, no. 5 (2013): 783–791.

Zhu, Liang. "Hypothermia Used in Medical Applications for Brain and Spinal Cord Injury Patients." *Advances in Experimental Medicine and Biology* 1097 (2018): 295–319.

# BIBLIOGRAPHY

## Deep Hypothermic Circulatory Arrest

Bigelow, Wilfred G., W. K. Lindsay, and W. F. Greenwood. "Hypothermia: Its Possible Role in Cardiac Surgery: An Investigation of Factors Governing Survival in Dogs at Low Body Temperatures." *Annals of Surgery* 132, no. 5 (1950): 849.

Chau, Katherine H., Bulat A. Ziganshin, and John A. Elefteriades. "Deep Hypothermic Circulatory Arrest: Real-Life Suspended Animation." *Progress in Cardiovascular Diseases* 56, no. 1 (2013): 81–91.

Gupta, Prity, et al. "Varying Evidence on Deep Hypothermic Circulatory Arrest in Thoracic Aortic Aneurysm Surgery." *Texas Heart Institute Journal* 45, no. 2 (2018): 70–75.

Shumway, Norman E. "C. Walton and F. John." *Annals of Thoracic Surgery* 68, no. 3 (1999): S34–S36.

## Emergency Preservation and Resuscitation

Bugos, Claire. "Doctors Put a Patient in Suspended Animation for the First Time." *Smithsonian*, November 22, 2019. www.smithsonianmag.com/smart-news/doctors-put-person-suspended-animation-first-time-180973626/.

"Crime Map & Data Stats," Baltimore Police Department. Accessed June 7, 2020. www.baltimorepolice.org/crime-stats/crime-map-data-stats.

Henig, Robin Marantz. "Surgeons to Put Gunshot Victims into Suspended Animation." *National Geographic*, April 2, 2014. www.nationalgeographic.com/news/2014/4/140402-suspended-animation-gunshot-victims-science-death/.

Murphy, Kate. "Killing a Patient to Save His Life." *New York Times*, June 9, 2014. www.nytimes.com/2014/06/10/health/a-chilling-medical-trial.html.

Robson, David. "The Ultimate Comeback: Bringing the Dead Back to Life." *BBC*, July 7, 2014. www.bbc.com/future/article/20140704-i-bring-the-dead-back-to-life.

Twilley, Nicola. "Can Hypothermia Save Gunshot Victims?" *New Yorker*, November 16, 2016. www.newyorker.com/magazine/2016/11/28/can-hypothermia-save-gunshot-victims.

Wu, Xianren, et al. "Induction of Profound Hypothermia for Emergency Preservation and Resuscitation Allows Intact Survival After Cardiac Arrest Resulting from Prolonged Lethal Hemorrhage and Trauma in Dogs." *Circulation* 113, no. 16 (2006): 1974–1982.

# INDEX

Abel, Renault, 161
absolute zero, 6
acidosis, 216
afterdrop, 46
afterlife, belief in, 125–126, 184
Air Force, US: suspended
    animation research, 111
alcohol: effect on hypothermia,
    19–20, 100–101
Alcor, 172–173, 177
Aldini, Giovanni, 141–144,
    222–223
Alexander, Leo, 76–77, 79–80
Allen, Frederick, 87–89
American Academy of Neurology,
    211
American Space Medicine
    Association, 101
amputative procedures, 87–88
Amundsen, Roald, 15
Andjus, Radoslav, 146–149
anesthetic value of cold, 27–29,
    87–88

aneurysm, near-death experience
    following, 186
animal research. *See* nonhuman
    animal research
animals, hibernation in, 92–93
anti-inflammatory response,
    56–57
Arctic regions: human habitation
    and survival, 14
Aristotle, 92–93
Armstrong, Neil, 112
arthritis, cold-water swimming
    relieving, 54–55
artificial cooling technology,
    71–72, 80–81, 140–141
"artificial hibernation," 85
artificial insemination, 155
artificial respiration, 139
Asimov, Isaac, 158
atomic motion, temperature and,
    6, 122, 129–131, 202
auditory memories, 187–189,
    191

243

embodied cognition, 198–199
emergency preservation and
	resuscitation (EPR), 217–220,
	224
empiricism, 35
energy output: human
	metabolism, 11–12
*Engines of Creation* (Drexler), 172
*An Enquiry into the Right Use and
	Abuses of the Hot, Cold and
	Temperate Baths in England*
	(Floyer), 43, 57–58
enzymatic function, 10–12, 84
ethics
	animal research, 146–150,
		223–224
	cryonics storage, 165–171
	head transplantation, 192–200
	informed consent, 85–86
	memory preservation, 205
	profound hypothermia
		treatment for traumatic
		injury, 218–219
	rat studies, 147–148
	using data from Nazi
		hypothermia experiments,
		78–79
Ettinger, Robert, 156–160, 171
evolution, mammalian, 118–119
experimental philosophy,
	128–129
extremophiles, 11

Fahrenheit, Daniel Gabriel,
	38–40
Fahy, Greg, 201–202

"fall of water" douches for
	psychiatric treatment, 60–65,
	81, 222
Fay, Temple, 70–76, 80–81, 84,
	86–87, 223
fever, 28–29, 47–49
Florentine thermoscopes, 37
Floyer, John, 42–44, 57–59
foods: association with cooling
	effects, 28–29
Forbes, William, 82–83
Forster, George, 142–143
*Frankenstein; or, the Modern
	Prometheus* (Shelley), 144
Freedman, Toby, 104–105
Freemasons, 39–40
freezing point of water, 11, 38–39,
	41
From Maggots to Murder (*Von
	den Maden zum Mörder*;
	Madea and Lignitz),
	100
frostbite, 13, 20, 29–30, 99–100,
	149
"frozen sleep," 73–74
fungal infections, 11

Gaia hypothesis, 148
galagos ("bush babies"), 149
Galen of Pergamon, 28–29, 36
Galilei, Galileo, 34
Galvani, Luigi, 141
galvanic stimulation, 141–144
gangrene, 87–88
*German Medical Weekly*,
	97–99

**PHIL JAEKL** is a cognitive neuroscientist and science writer. Aside from his academic publications, he has written on topics related to neuroscience for popular periodicals such as *The Atlantic, The Guardian,* and *Wired* magazine. He has also written for *New York* magazine and has contributed feature-length essays to *Aeon.* What motivates him is the desire to bridge fascinating yet complex scientific breakthroughs— often only covered in subscription journals—with explanations and experiences that make the discoveries relevant and meaningful to as many as possible. His goal is to be a communicator of science.

This work was inspired by an essay previously published in *Aeon* magazine called "In Cold Blood."

Phil grew up in small-town southern Ontario, Canada, where he went on to study psychology in Toronto. He completed graduate studies at York University, specializing in the Brain Behaviour and Cognitive Science program, through which he earned a PhD. After completing postdoctoral research in Barcelona, Spain, and in Rochester, New York, concerning how the brain combines information from different senses, he and his wife moved to Tromsø, Norway. His postdoctoral work on how the brain "hears" distance has been covered in *Scientific American.*

From where they reside at about 70 degrees latitude in the High Arctic, Phil and his wife enjoy all manner of outdoor activities, ranging from mountain hiking and snowboarding to photography and fishing. Equally enjoyable to Phil is the sensation of warmth, whether it comes from the sun, a fire, or a tasty meal.

PublicAffairs is a publishing house founded in 1997. It is a tribute to the standards, values, and flair of three persons who have served as mentors to countless reporters, writers, editors, and book people of all kinds, including me.

I. F. STONE, proprietor of *I. F. Stone's Weekly*, combined a commitment to the First Amendment with entrepreneurial zeal and reporting skill and became one of the great independent journalists in American history. At the age of eighty, Izzy published *The Trial of Socrates*, which was a national bestseller. He wrote the book after he taught himself ancient Greek.

BENJAMIN C. BRADLEE was for nearly thirty years the charismatic editorial leader of *The Washington Post*. It was Ben who gave the *Post* the range and courage to pursue such historic issues as Watergate. He supported his reporters with a tenacity that made them fearless and it is no accident that so many became authors of influential, best-selling books.

ROBERT L. BERNSTEIN, the chief executive of Random House for more than a quarter century, guided one of the nation's premier publishing houses. Bob was personally responsible for many books of political dissent and argument that challenged tyranny around the globe. He is also the founder and longtime chair of Human Rights Watch, one of the most respected human rights organizations in the world.

$\cdot$     $\cdot$     $\cdot$

For fifty years, the banner of Public Affairs Press was carried by its owner Morris B. Schnapper, who published Gandhi, Nasser, Toynbee, Truman, and about 1,500 other authors. In 1983, Schnapper was described by *The Washington Post* as "a redoubtable gadfly." His legacy will endure in the books to come.

Peter Osnos, *Founder*